Designing Your Car

As you design and build your race car, you will face several critical crossroads and decisions, such as which material should I use here, will that be strong enough, and will it have enough clearance there. This chapter is written to give you a general philosophy which should guide you through those agonizing decisions with the least worry, and give you a quality race car as the finished product.

CHOOSING THE BODY STYLE

There are several factors to weigh in choosing which body style to use for a race car, but the most important factor before any others are considered is the availability of the body in wrecking yards and the price of the replacement sheetmetal. If the body and sheetmetal are scarce, or the cost is higher than other popular styles, you will be agonized by the frustration of trying to purchase all the parts you need throughout a racing season.

The other factors relating to the body style choice are total body frontal area, body weight, and wheelbase and overall length.

The frontal area is not just the grille and headlight doors, but also the windshield area and any other portion of the body which meets the oncoming wind head-on. For short tracks and slower speeds, the body frontal area is of minor importance, but as speeds increase, it gets more important. On a half-mile or larger track, it does start to gain in importance.

The body weight is determined by the amount of sheetmetal area a particular body uses. Many cars with the same wheelbase use different amounts of sheetmetal, thus yielding a variance in body weights. This becomes important even if you have to meet a minimum weight requirement. The body should be as light as possible, as should the entire car, allowing the finished race car to weigh less than the required minimum. This will allow you to add ballast weight in more desirable places in order to meet the minimum weight requirement.

Wheelbase and overall length of a car are important in considering the maneuverability of a race car through traffic. A smaller car will maneuver better. A shorter wheelbase is also more sensitive to weight changes. In other words, it is easier to shift more weight distribution to the rear with a shorter wheelbase. Also related to this matter is track width. For oval track racing, as wide a track width as is allowed is desirable to help limit inside-to-outside weight transfer during cornering.

CHOOSING AN ENGINE SIZE

The classic choice is between the big block and the small block engines. Currently it costs more to produce the same amount of horsepower out of a small block than a big block, and small block durability is not as great. However, that will be changing over the next couple of years as big block engines get phased out of production. Parts will become harder to get, the big blocks will be used less and less for high performance, and it will become increasingly difficult and costly to get speed parts for the big blocks.

The Camaro and the Nova are both very popular body styles for racing because they offer wide availability of replacement sheetmetal and also fiberglass body parts. While the Camaro is slicker in the air than the Nova, both offer some aerodynamic advantages. Both are also compact in size for their wheelbase, offering favorable weight advantages.

The biggest disadvantage of the big block engines is their weight penalty. Most big blocks weigh roughly 100 pounds more than small blocks, and that is a hefty amount of extra weight to be carrying around up front in a racing vehicle.

You should consider this advantage of small blocks and light weight: tires are now designed for the ligher cars. This is because the testing the tire companies do is with the most competitive cars, which are the lighter cars.

WEIGHT DISTRIBUTION

Through your entire race car building process, always keep uppermost in your mind the goal of getting as much static weight as possible off the front end. Remove the weight altogether if possible, or shift it further back along the wheelbase line if possible. Gaining a greater rear weight bias is going to help the problem of power oversteer of a car coming out of a turn. With wheels under hard acceleration exiting a corner, the forward traction capability of the tires is going to steal side traction ability away from the tires. The problem is complicated by the fact that the rear springs must be stiffer to balance a nose heavy car. The result is oversteer. Additional weight on the rear wheels will shift more roll couple distribution to the front springs, and add to the traction ability and side bite ability of the rear tires.

ENGINE SETBACK

Whenever allowed by the rules, set the engine back in the chassis as much as possible. With the engine set way back

such as in a modified where the car almost becomes a mid-engine vehicle, the center of gravity is not displaced very much under accelerating and braking. This is especially important under braking, where as little forward weight transfer as possible is desired. This allows all four wheels to handle the braking effort more equally and allows the vehicle to drive deeper into the corners. The further back the engine sets in the chassis, the more these desirable handling characteristics will effect the chassis.

WEIGHT SAVING

Tied hand-in-hand with the favorable weight distribution characteristics is the factor of saving all the weight you can. So often we have heard racers say, "Oh, I only had the big tubing on hand so I made it out of that. It's so small the weight doesn't make any difference anyway." If the car crafter carelessly adds just small amounts of unneeded weight in places where "it doesn't make any difference anyway", he could well add as much as 100 to 250 extra pounds to his car without even realizing it.

Every bit of weight counts, no matter how small. Remember that if you removed an ounce of weight in 160 different places on your car, you would save 10 pounds. Usually the choices you are faced with involve more than an ounce, so you can see that there are opportunities to make considerable weight savings. Every time you add a piece to your race car, think to yourself, "Is there a way I can make this piece lighter without sacrificing safety or structural

Race Car Fabrication & Preparation

by
Steve Smith

Editor	Steve Smith
Associate Editor	Georgiann Smith
Production Assistant	Lori Larson
Drawings By	Tim Forhan and Steve Smith

Copyright © 1977 by Steve Smith Autosports. No part of this publication may be reproduced in any form or by any means without the express written permission of Steve Smith Autosports.

ISBN No. 0-936834-14-5

Published by

Steve Smith Autosports

P.O. Box 11631, Santa Ana, CA 92711

Table of Contents

THANKS

No book of this size and magnitude could ever be written without the help of hundreds of people. We would like to extend a special thanks to Tex Powell, Frank Deiny, Ivan Baldwin, Sonny Easley, Dave West and Ken Sapper Jr. for the help they provided. Also a big thank you goes to the many people who answered questions and dug for information for us. And most importantly, thanks to a very understanding and dedicated wife who put up with me while this project became all-encompassing.

Steve Smith

Engine setback is one of the easiest ways to gain the desirable rear weight distribution, if your racing association allows it.

chassis and not the springs is absorbing the loads imposed by cornering, and this is to be avoided. To steer clear of these problems, careful planning must be done in chassis and component layout which allows maximum travel of the wheels without any component bottoming. At least six inches of upward wheel travel on the outside should be planned for.

RULES INTERPRETATION

The first step in building a race car is being thoroughly familiar with the rules of the division and association under which the car will be raced. The obvious rules take care of themselves, but there are always some hazy areas which come up during construction. These are details which are not covered in the rule book, or which raise a question in your mind. The best policy is to interpret any shadow areas to your advantage. If the technical inspector should interpret the rule book differently than you did, then let him find the infraction and ask you to change it. Of course, we are talking about small areas here, like bumper heighth clearance, fender well shaping and clearance, or substituting Lexan plastic for a glass windshield. We do not suggest or encourage flagrant rule violations.

One caution: do not take advantage of, or bend, safety rules. They are designed to protect you and anybody else who may drive the car, now or in the future.

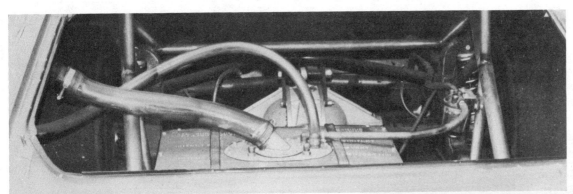

Note that the fuel cell is situated as far to the left and as far forward as possible.

integrity?''

Other examples of weight saving: use Lexan plastic for the windshield (it has better qualities and is safer than plexiglass), and use aluminum or magnesium for every non-stressed metal item such as driveshaft loops, window clips, etc. Also, you can save weight by cutting down many required items such as bumpers.

BODY ROLL CLEARANCE

The trend today is toward softer springs in race cars, which give the advantage of more traction. Softer springs produce more body roll, however, so how soft you can equip your car is limited by how much roll clearance is available. As the body rolls over, something is going to hit — shocks absorbers bottom, wheels bottom on the fenders, A-arms bottom on the frame, coil springs bind, suspension linkage bottoms on the chassis, or the bottom of the car drags the track. As soon as one of these events occur, **the**

The battery placed in a secure and protected position just in front of the left rear wheel aids greatly in favorable weight distribution.

ENGINE	CID	OIL SUMP (stock)	STARTER	WEIGHT	A	B	C	D	E	F	G
American Motors 6	232/258	rear	left	525	24"	24	29	35	26	30	25
American Motors V8	304/360/401	rear	right	540	21½"	25½	29½	29¼	28	28¾	21½
Buick V6		middle	right	370	24"	26	28½	23	25½	23	21½
Buick V8	350	rear	right	450	23	28½	28½	30½	25½	29	21½
Buick V8	430/455	middle	left	600	23	28	30	30	27	29	22
Chevy 4	153	front	right	350	16	23½	28	23½	25½	22½	24
Chevy/Pontiac Olds/Buick 6	250	rear	right	410	16	23½	28	32½	25½	30½	24
Chevy V8	265/283/302 307/350/400	rear	right	550	19½	26	27	27	25	26½	20½
Chevy V8	396/402/427 454	rear	right	625	22	27	33	30½	29½	30½	23½
Ford 6	144/250	front	right	400	17	17	28	31	26	29	24
Ford 6	240/300	front	right	400	13	13	28	32	26	30	24
Ford V8	260/289/302 351	front	right	460	20	22	27	29	25	27	22
Ford V8	332/352/390 410/428	front	right	625	23	27	32	32	30	30	28
MoPar 6	170/198/225	front	right	475	24	30	26	31½	22	30	29½
MoPar V8	273/318/340 360	front	left	550	20½	25	31	29½	28	29½	23½
MoPar V8	383/400/440	front	left	670	23½	29½	30½	30	28	29	24
MoPar Hemi	426	center	left	690	28½	29	31	31	28	32	24
Olds V8	330/350/455	rear	left	600	21½	26½	31	31	27	29	24
Pontiac V8	350/400	rear	left	590	22	27	31	29	26	28½	20
Pontiac V8	455	rear	right	640	23	27	33	32	28½	29½	27
Cadillac V8	472/500	middle	right	600	23½	28	32	30½	29	30	28½

If you are considering the weight and size advantages of any particular engine, this chart and the drawings can help answer any of your questions.

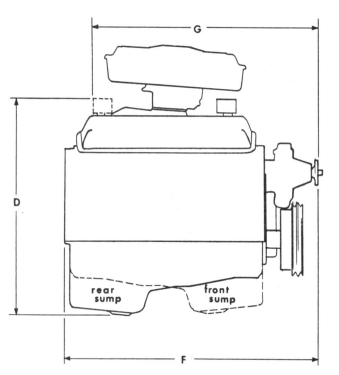

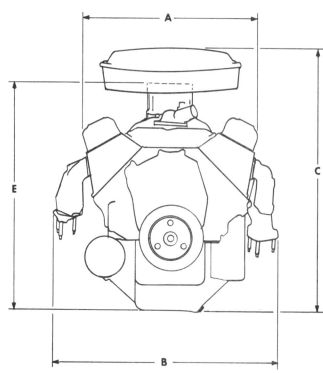

The Safety System

Racing is risky—even at slow speeds. There have been serious injuries and death on short tracks where speeds are relatively slow and bumps are relatively mild. Enough said. It is the responsibility of the car builder to build in all the safety he can—even if he himself will be the driver. This should be done even if it costs extra weight and money. Safety is an area where shortcuts don't pay. We'll explore the important areas of safety in a racing vehicle to make sure you get the most out of your safety systems.

SAFETY HARNESSES

It just is not enough to install a seat belt and shoulder harness. The selection, installation and maintenance of these lifesaving devices is important if maximum protection is going to be afforded the driver.

The first consideration is the quality of the belt itself. Belts and harnesses are made of nylon webbing, which can vary in its tensile strength by material width and thickness. When purchasing a new set of belts and harnesses, check the tensile strength rating and purchase the ones which have the greatest strength. Then, treat these belts with tender loving care. A recent test carried out by a testing laboratory showed surprising results about the effects of the elements found inside a race car cockpit on the life expectancy of nylon belts. The lab did a tensile strength test on a standard 2-inch wide shoulder harness which had an original tensile strength rating of 6,600 pounds. After a nine month usage in

The shoulder harness is anchored to a bracket welded to a roll cage brace to provide maximum strength and safety. All metal the harness passes over is padded to prevent chafing the material.

This happened after a stuck throttle on a half-mile dirt track. After a driver walks away from a nasty bump like this, he is mighty thankful he adhered to strict safety requirements.

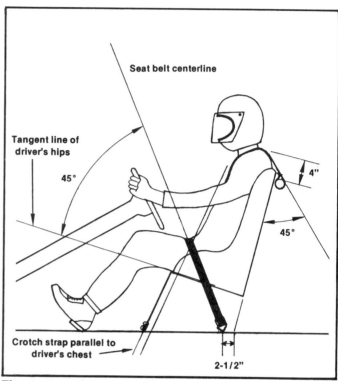

The text explains why these safety harness mounting positions and angles are vitally important. Pay attention to them — they could prevent injuries.

a race car where it had been subjected to sunlight, sweat, grease and dirt, the tensile strength had dropped to 1,600 pounds! An incredible loss of strength! Think what the strength is of belts which have been in continuous service for two racing seasons! The lesson to be learned here is that belts and harnesses should be treated with care, and replaced at the first sign of abrasion, fading or wear, or definitely once every other season.

The mounting hardware for the belts is just as important as the belt quality. It doesn't make sense to have high quality belts mounted with hardware that will bend, break or pull loose under light impact. The buckle latch is the key to the entire safety harness system. Presently there are two types of latches which have been standardized by the safety harness manufacturers: the quick release and the hook latch. The hook latch is by far the least expensive, and it does its job perfectly. The latch-type of fastener is made with the harness ends fitting over a slotted tongue which is secured by a locked lever. The lever is held in place by a ball-and-detent mechanism. The entire latching mechanism should be checked periodically for wear and the presence of dirt or foreign materials which could harm its operation. Remember that the buckle must lock securely and open quickly. The ball-and-detent mechanism wears through normal usage, and when it does, may prevent the latch from locking positively or may cause premature release. When it latches or releases quite easily, let it be a warning to you that it is worn out and should be repaired or replaced.

The mounting hardware should be a minimum of grade 5 hardware in 3/8-inch or larger diameter. All eye bolts, shoulder bolts or other securing hardware should have a minimum tensile strength of 50,000 pounds. Eye bolts

should be one-piece forged with a minimum of 3/8-inch shank, and not a formed and welded hook.

The mounting positions and angles of the belts are very important, in terms of how the belts place a load on the mounting hardware (it has less strength when stressed at an angle instead of straight-on) and how the driver's body absorbs rapid deceleration or impact forces. The seat belts should be mounted at an angle of 45 to 55 degrees from a tangent line of the thighs, according to a U.S. Government study, and their anchoring point at the bottom of the seat should be 2-1/2 inches forward of the back line of the driver (see drawing for clarification). Also of importance is the distance between the bottom anchoring points for the belts. The distance between them should be the same as the width of the driver (anything wider stresses the belts at an angle rather than straight on).

The shoulder harness attachment angle should be at approximately 45 degrees from the driver's spinal axis. In addition, they should anchor to a bracket on a roll cage cross brace about four inches below the driver's shoulder line. This is of utmost importance because if the straps are mounted too far above the shoulders, they will allow the driver to move upward in the seat before being restrained. If they are anchored too low below the driver's shoulders, they could compress the driver's spine during a sudden downward jerk.

It is also very important to use a crotch strap or antisubmarine strap with the safety harnesses. It secures the belt across the driver's hips in the event of an accident so that it is not pulled up to the stomach or lower rib area by the shoulder straps. It also prevents the driver from sliding forward and under the belt. To be most effective, the crotch strap should be anchored in line with the chest of the driver as seen from the side view. If the crotch strap is mounted too far forward of the driver, unnecessary injury could result because his body will slide partially out of the seat before he contacts the belt and is then restrained.

The length adjusters for the shoulder harnesses should be placed at a line parallel with the driver's armpits. If they are allowed to be placed any higher, the adjusters could seriously bruise or crack the collar bones in the event of an impact.

Seat belts and shoulder harnesses should pass from their anchoring points directly to the driver. They should not be allowed to drape over the seat sides or back, but rather they should pass through access holes cut into the seat. If the belts draped over the seat, during an impact stress the belts would deflect the seat leaving the driver virtually unprotected. If the seat were a fiberglass seat, the belts most probably would break the seat.

When you tighten your harnesses, tighten the seat belt first, then tighten the shoulder harness. This prevents your pulling the seat belt up around the rib cage area. The antisubmarine strap would also help prevent this.

All safety belts should be adjusted super tight before each race because perspiration and driver jounce allow the belts to loosen up as a race progresses.

SEATS AND SEAT BRACING

The primary function of the roll cage is to protect the driver in case of an impact or roll-over. Because the roll cage is supposed to remain intact even though the body structure may disintegrate, it is only logical that the driver should remain securely located within the roll cage structure. To accomplish this goal, the seat should be attached to a tube frame which is securely attached to the roll cage—and definitely not to the floor pan.

The frame which holds the seat and is welded to the roll cage must be designed to handle forces fed into the seat from an impact evenly, and the seat and frame should be adequate to withstand 20 G's of force. Use 1-1/4-inch O.D., .125-inch wall round tubing to form the seat mounting (see drawing for more details). The plates and brackets which are welded to the seat mounting frame should have a minimum thickness of 3/16-inch, and care should be exercised not to drill the mounting holes too close to the edge of the bracket, so a fatigue crack from the hole cannot form. The seat frame and seat should be positioned as close to the floor as possible to get that big piece of ballast (the driver!) as low in the car as possible.

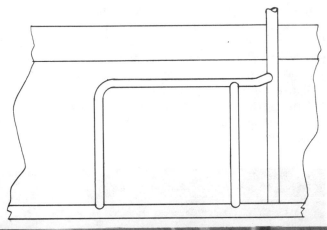

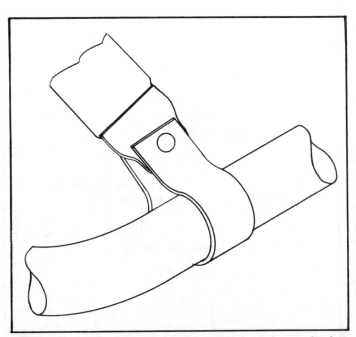

Where a seat belt welded bracket might be sub-standard, a steel strap wrap-around bracket like this one can provide the needed strength. The bracket is made from 1-inch by 1/8-inch thick steel strap rapped around the cage seat rail. This method reduces the possibility of bracket bending and places the bolt in double shear, which is good.

Drawing [top] shows seat mounting frame in plan view, and in photo above. The seat frame attaches to the bottom door bar at its left and the roll cage at the rear and not to the floor. This way the seat and driver will move to the right when another car crashes into the left door.

With a fiberglass or plastic seat, use 3/8-inch O.D., .065-inch wall round tubing to form a supporting frame around the complete perimeter of the seat, across the back and down the center of it. This frame gives secondary support for the driver should an impact ever shatter the fiberglass seat. It also provides extra support against flexing for the heavier driver.

The seat is mounted at six positions on the roll cage tubing frame, four on the bottom of the seat and two at the upper back of the seat. Be sure to use rounded head bolts to insure driver comfort for mounting the seat, with a large washer placed under each bolt head to prevent the bolts from pulling through the seat, or work-fatiguing the seat.

Any supporting brace holding the seat in its upper back region should have a rounded edge facing the driver's back. Pieces of tubing or rod facing the driver could leave two nice punctures in the driver after he has backed the car into the third turn wall!

The seat itself should be of high quality construction, and above all, comfortable for the driver to sit in. If he is uncomfortable or trying to hold himself up, it is difficult for him to concentrate on driving a good race. The seat should be designed to support the driver laterally against cornering forces, but not with a rigid mount or brace which could break a driver's hip or ribs in case of a side impact. We have actually observed this mistake of solid bracing carried to the extreme of having a piece of tubing running from the floorpan to the seat. The seat should support the driver **evenly** at his thighs, hips, ribs and shoulders. The total support should not be taken at the ribs. A properly designed, quality seat is the answer.

A padded head rest should be installed behind the driver's head, no more than four inches behind his helmet so he is protected from whiplash injury.

Note the sturdy-but-giveable headrest, and the good uniform lateral side support of the seat.

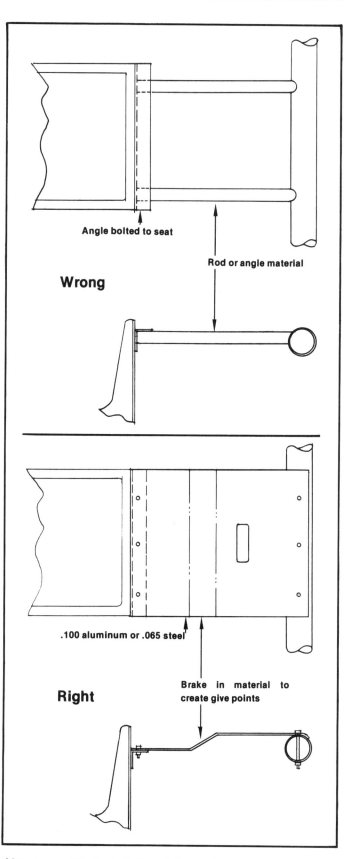

Angle bolted to seat

Rod or angle material

Wrong

.100 aluminum or .065 steel

Right

Brake in material to create give points

Above are illustrated the right and the wrong ways to mount the upper support of the seat to the roll cage. The right way allows for a flexing of the support in case of a hard impact.

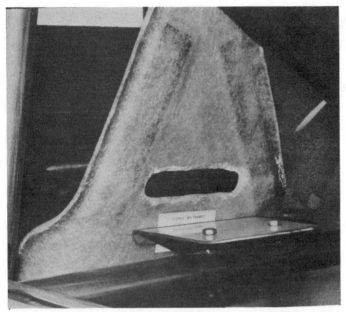

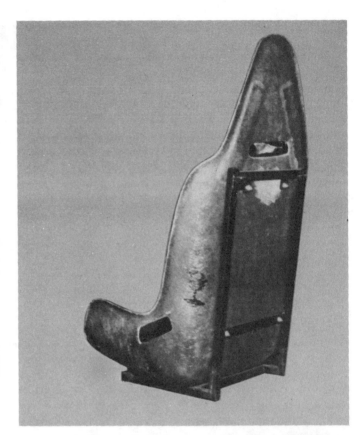

Above is shown a seat mounting which is partially correct. Aluminum plate with a bend at the front will protect driver, but material should have a brake in it. At right is one method of constructing a fiberglass seat bracing which will bolt to roll cage seat frame.

All tubes within reach of the driver's body and head should be padded with a rigid, dense foam—not a soft, flexible foam—to absorb impact should he ever contact the tubes. Under impact, the soft foams will bottom quickly and allow the helmet or body to contact the rigid surface. A widely used roll cage foam is Armstrong Armaflex rubber air conditioning insulation. Remember that foam deteriorates, so new protective foam should be installed at the beginning of each season.

THE FIREWALL

As the name implies, it is a wall which seals the driver's cockpit from the engine to protect against fire. Be sure all holes in the firewall are sealed. If a fire starts while the car is at speed, a small hole will act just like a torch with the air forcing the fire through it.

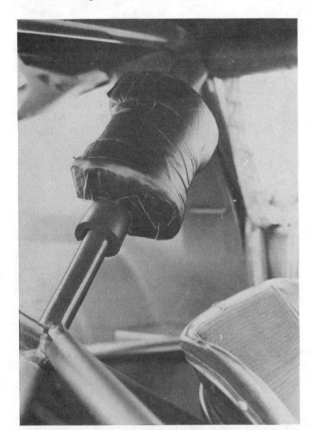

Above is a simple headrest made from rigid foam and black tape. At right, a stock firewall is too heavy and massive for a race car. It is also full of holes which must all be plugged. Best bet is to fabricate one from sheet steel and weld it in place.

Very nice simple headrest made from .065 sheetmetal, with two brakes in it to allow a give.

Another illustration of the roll cage seat frame, with a fiberglass seat bolted to it. Aluminum plate in bottom of seat prevents stress concentrations around bolt heads which could tear the fiberglass.

It is mandatory for driver protection to have the center of the steering wheel lined with a rigid foam. The steering wheel should be covered with some type of material which will not get slick when driver's hands perspire. Silver tape on top of wheel shows driver where straight-ahead position of wheel is.

The offset steering shaft will fold up in two different directions around the U-joints if the steering shaft receives a head-on impact. The aluminum stabilizing block would be torn loose by impact.

STEERING COLUMN

So that the driver does not get speared with a steering column or steering wheel, use a collapsible column, or an offset shaft. The collapsible steering columns are found on all late model passenger cars. The offset steering shaft can be constructed by using two universal joints in the shaft, setting the three pieces of the shaft in different planes. Dished steering wheels are desirable, with a piece of rigid foam taped to the center of it.

If you use a custom steering wheel with metal spokes, make sure they are steel — not aluminum — and mount to a sturdy steel bracket on the steering column. A driver could easily tear the aluminum material out and leave him without a steering wheel — it's happened!

FIRE EXTINGUISHING SYSTEM

If you are planning to use an on-board fire extinguishing system, here is a checklist for you: 1) Use only systems which employ Halon 1301 (previously called Freon 1301) as the fire extinguishing agent. There are other types of Freon, but they can be lethal in a fire. Don't use a dry powder agent. 2) Steel bottles are better than aluminum bottles. The steel ones are refillable, whereas the aluminum ones have to be tossed (Department of Transportation ruling). 3) Use only a system which completely discharges upon activation. Who knows when a partial discharge is adequate, and when it is not? 4) Place the bottle in the car in a position where it cannot be damaged by collision or impact. Mount the bottle securely so it cannot come loose and fly around the car. The bottle should also be protected from excessively high heat. 5) Don't ever connect two bottles together into a common distribution system. One bottle could discharge into the other and explode it. 6) Periodically weigh the bottle to determine if it is fully charged. A pressure gauge does not indicate how much agent is in the bottle. Compare the weight of the bottle against the full weight shown on the label. 7) Remember that the only real test of the system is when it is really needed. Then proper maintenance can make a difference if your neck is saved!

The fire extinguishing systems cost money, but you can make a less expensive one yourself. Use two five-pound bottles. Place one shooting into the trunk, and one shooting into the engine compartment. Attach a cable to each and run it up to the driver. Keep the pin in the handle of each

bottle until the race starts so the bottles aren't accidentally set off. Then, don't forget to remove the pins.

Note the sturdy mounts of the fire extinguisher bottle. Also note how safety pin is retained while car is being serviced.

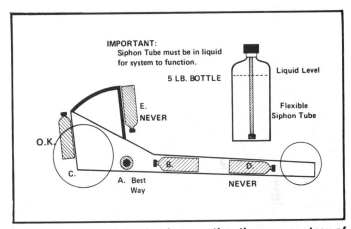

These fire extinguisher bottle mounting tips are courtesy of Specialty Equipment Manufacturer's Association.

WINDOW NETS

A good quality window net is important for driver protection even if your racing association does not require them. The net should release from the top so it will fall down out of the way when it is released. If the net releases from the bottom, it naturally hangs over the window opening and must be pushed out on top of the car for the driver to exit. In a panic situation, the driver may be slowed in his exit from the car while trying to fight a net hanging from the top.

OTHER SAFETY PRECAUTIONS

If the battery is mounted in the driver area, seal it in a plastic marine battery box, and provide amply sized brackets

The window net should mount so that it is retained at the bottom, not the top. In an emergency, gravity will put the net down out of a driver's way as soon as the spring latch at top is released.

and hardware to hold it in place. The battery is so dense it can fatigue and break incredibly strong mounts.

Beware of protrusions in the driver area that could harm the driver in a bad crash, such as a mirror, the corner of the battery box, shifter, shifter support box, switches and upper steering column supports and bolts.

Fuel pressure gauge hoses should be protected from electrical wires that could short out and burn a hole in the hose. Also protect the fuel hose from any sharp objects. The fuel pump can pump a lot of fuel in the cockpit if there is even a small hole in this hose.

The Electrical System

Just the thought of electricity and wiring makes some people uneasy. It can be a mystery area, and an area that can cause some pretty disparaging (but so simple) failures. With a little background information, however, it's simple.

ELECTRICAL SYSTEM BASICS

There are three terms associated with the electrical system which you should be familiar with in order to choose your components and wiring correctly for the job: volts, amps and resistance.

If an electrical system could be compared to a fluid system for visualization purposes, then **volts** or **voltage** would be compared to the pressure in a fluid system. Voltage moves the electric current through the circuit. It is the pushing force. **Amps, amperes** or **amperage** is just like the rate of flow in a fluid system. It is the volume or amount of the electrical current. **Resistance** in a circuit, measured in units called **ohms**, is how the flow rate is slowed down by the load. Resistance can be created by too small a wire gauge trying to carry too big a load.

WIRING THE SYSTEM

Electrical systems must be wired in a closed loop so that electricity flows back to its source. If the closed loop is opened to the atmosphere, there is an electrical leak (called a short). This can be a cut wire or a switch turned off. If there is an inadvertent short in the circuit, the electricity has little resistance (resistance is provided in the load, or in other words, the appliance the electricity is operating), and the battery discharges very fast. Because the fast travel of electricity causes heat in the wiring, a short can often times cause burnt wiring or a fire.

In a negative ground system (as found in almost every car made today), the negative terminal of the battery is wired to the full metal chassis of the car, making the entire chassis the negative terminal of the battery. So, the load (any appliance such as the starter motor, cooler motors, etc.) need only be grounded to the chassis instead of requiring a separate return wire for each load to complete the closed circuit back to the source. When completing your grounding of electrical appliances in the car, be sure they are **properly grounded**, meaning that the negative ground makes good contact with a part of the chassis which flows electricity

The closed loop type of spade lug, left, should always be used for any type of wiring connection in a race car. The U-type at the right can work loose from vibration.

back to the battery source. Improper grounding is a big cause of electrical system failures in race cars.

Wiring provides the pipeline for the electricity to flow through, and the first consideration is the size of that pipe. Wire is measured in gauge, with 00 being a very thick wire and heavy gauge, and 16 being a very thin wire and light gauge. For race car wiring, 14 gauge should be considered a minimum size. Consult the wiring diagrams in this chapter for a guideline on which wire gauge to use where. When in doubt as to the gauge of wire to use for an application, opt for the larger diameter as it can carry a bigger load without heating up.

Color coded wiring can be useful to you in wiring your race car, IF you keep records of which wire color goes where. If you use a terminal board with good identification on it, it will make for an easy system to maintain and repair.

A highly recommended brand of wire for a race car is Bowman 23437. Its insulation will not burn, but in case of extreme heat, the insulation will become molten. When the wire cools down, though, the insulation will form back around the wire.

Wiring terminal ends for a race car application should be of the ring type (a closed circle) which slips over a threaded post or is attached with a screw, rather than using the open-ended spade lug type of terminal. This prevents accidental vibration-loosening of a wire. Wires on studs should be secured with lock nuts.

For safety, wire terminal ends should be both crimped and soldered onto the wire. There shouldn't be such a thing as a "solderless terminal" in a race car. To insure against accidental grounding, vibration-loosening and vibration-caused metal fatigue of terminals, seal the terminal end/post connection with a small amount of rubber silicone (called "potting").

As a further insurance measure, use shrink tubing to protect soldered terminal/wire joints from the elements. Just slide a short length of the tubing over the joint (or any crimped splice you've made) and apply heat (like from a match or hair dryer). The idea is to shrink the tubing without melting it, so be sure to keep the heat source quite a distance away. Like magic, the tubing will shrink down to about half its orignal diameter, and tightly encase the electrical connection. This shrink tubing is readily available at electronics supply stores.

In choosing terminal ends for wires, remember there is one size of terminal for 14 gauge wire, another for 10 gauge, another for 6 gauge, etc. A 14 gauge wire slid inside of a 10 gauge terminal just won't work too long.

Routing your wire is important to the overall reliability of the electrical system. Never route the wiring over or through anything sharp without using a rubber grommet protection to line the hole. Short lengths of rubber hose can be substituted for grommets in a pinch. Vibration will cause the wire to rub against the sharp object and eventually chafe the insulation, which will probably then mean a short. Wiring should also not be routed near any high heat components.

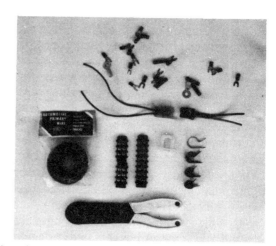

Local parts stores have a wide variety of wiring tools and hardware available to solve any problem you may have.

An auxiliary "black box" is a very common item in cars using electronic ignition. The quick-disconnect wiring allows a rapid change from one box to the other. Note the stud-type terminal board with color-coded wiring, and the taped wiring which prevents chafing of wires.

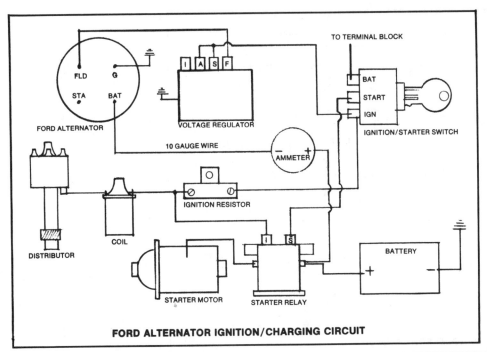

FORD ALTERNATOR IGNITION/CHARGING CIRCUIT

Whether you are racing a Ford or GM product, the basic wiring required is very simple and straightforward.

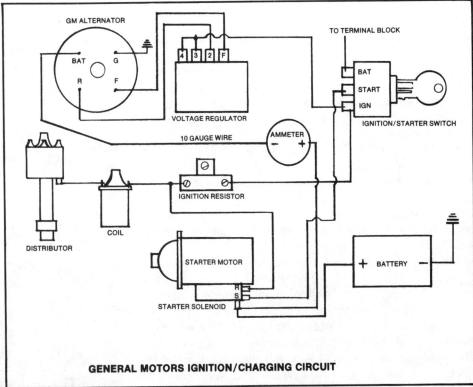

GENERAL MOTORS IGNITION/CHARGING CIRCUIT

Most standard wiring has a PVC plastic insulation, which is easily affected by high heat. There are silicone and teflon wire insulation coatings for wire which can be used if a high heat situation is encountered. Do not route wires near fuel or oil lines because in the event of a short, the wire could burn a hole in the line.

On a race car, you are constantly bolting and unbolting certain electrical components. For these items, quick-disconnect plugs help speed up your operation. Use them in places like distributor primary lead, alternator and starter solenoid. They will make engine changes a lot easier. These types of plugs are frequently used in the under-the-dash region of most late model passenger cars. They are also available from trailer supply stores, and some electronics supply stores.

To make your wiring chore easier and to separate one electrical circuit from another, use a bakelite (a hard black plastic) terminal strip which can be purchased cheaply from

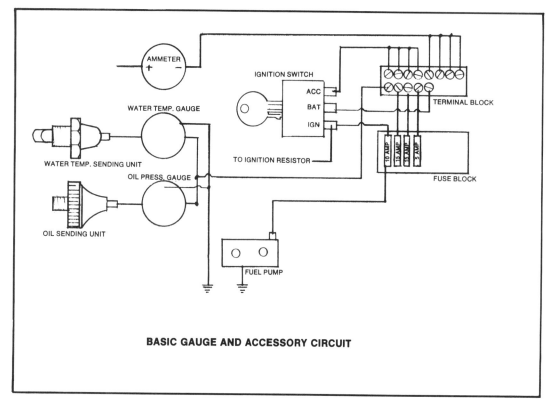

BASIC GAUGE AND ACCESSORY CIRCUIT

an electronics supply store. It is much better to keep each type of electrical circuit separate from one another to provide overload protection, and to provide ease in tracing failures. Be sure to mount the terminal strip in an easy-to-repair location.

Tired of carrying a box full of various amp fuses around in your tool box? Instead of using fuses, use automotive circuit breakers. This way, once you repair the overload condition, power can be restored without extra cost or time. The circuit breakers are more durable for race car installation also. They are available in a variety of amp ratings at trailer supply stores or auto parts stores.

The amp ratings of the breakers (or fuses, if you use them) should be carefully chosen. If you use a breaker rated at far too many amps than the circuit is carrying, then a circuit problem is never going to trip the breaker or burn the fuse, and you run the risk of burning up part of the electrical system, or at least ruining the problem appliance. You should always use a 5-amp fuse in a 5-amp circuit, and then if it burns, you know you have a problem to solve.

Switches, also, should be matched for amperage. When you install a switch be sure its amperage rating matches or exceeds the maximum amperage of the circuit. A 5-amp switch controlling a 10-amp circuit is likely to burn out quickly. Or, it might continually blow the fuse or circuit breaker without your ever discovering the cause of the problem. Beware of cheap switches. They can vibrate apart, sometimes during a race, allowing a hot wire to fall on a metal part and short out.

In choosing switches, you have two types of terminals to

The electrical components employed in a race car are so few and basic, that a wiring job can be considered a "minor" job. The main point here is to be sure to use the proper size fuses or circuit breakers, and keep the wiring coded and mapped for quick service or repairs.

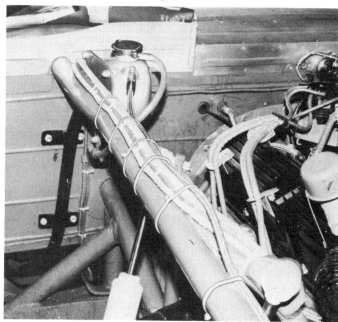

To make a very neat and professional appearance under the hood, tie-wrap all the wiring and plumbing lines together to a roll cage tube. Note the rubber grommets where anything passes through the firewall. Also note the 9 millimeter plug wires and their tie-down brackets.

look at: soldered and screw connector. A soldered joint is much better since it resists vibration, a constant enemy. On the other hand, it takes a little longer to replace than a screw connector, but if you start with high quality switches (and the proper amp rating), you won't be making many switch changes anyway.

HARNESSING YOUR WIRES

Keeping everything in place keeps things looking neat. It also prevents problems from wire chafing, burnt or accidentally cut wiring, wires getting in the way at inopportune times, as well as fatigue cracks in the wiring. For wiring running a distance together, bundle it in clear plastic wire wrap, available in several tube diameters from electronics supply stores. Or, shrink tubing can be used to waterproof and protect the wires and give the job a neat, professional look. Remember to use nylon tie-wraps and/or rubber-cushioned Adel clamps to keep wiring adequately retained in its place.

The rubber-lined Adel clip protects wiring and tubing. The cushion assures tight fit, eliminates vibration.

IGNITION COMPONENTS

Electronic and transistorized ignition components are sensitive to overdoses of heat and vibration so care should be taken to isolate them from these destructive elements.

The electronic ignition control box as well as the starter solenoid (remote type) and voltage regulator should be mounted on an aluminum board located inside the car (usually on the floor, at right front). If something goes wrong at the race track (where else would it?), you won't spend a lot of time diagnosing the problem. Just take out the module board and replace it with a spare one. Use plug-in wiring connectors (available from trailer supply stores) on all the wires. Mount the board with soft rubber insulators under each bolt to prevent vibration damage to the components.

Spark plug wires should be positioned through non-conductive wire mounts to prevent their contact with conductive metals or hot surfaces. It also prevents wires from contacting each other which could cause cross firing. Use nine millimeter spark plug wires to stem voltage leaks. The wires should have silicone rubber covering which makes them impervious to water, oil, chemicals and heat.

DISCONNECTING

In SCCA racing, a quick-disconnect switch (or master kill switch) is required in the positive battery terminal wire to quickly deactivate the entire electrical system. If you are using an electric fuel pump in your car, it is strongly recommended that you use one of these in an easy-to-reach spot. It is also handy to have when you want to quickly recharge the battery and don't have time to yank off a battery cable. Electric fuel pumps should be wired to a mercury switch which switches them off should the race car roll over.

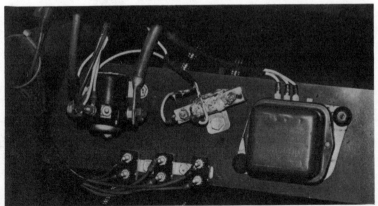

Left and above, two more ideas for ignition components mounting boards inside a race car. Both mount the Ford solenoid and ballast resistor inside, both for convenience and protection from heat.

THE STARTER

On Chevrolet small block engines, the starter is located on the lower right hand side of the engine. This presents a problem as the optimum mounting position for the battery in an oval track car is at the lower left hand corner of the firewall. The Chevy starter has its solenoid located on top of it, and requires a hot lead from the positive side of the battery. This would entail running a wire across the engine in the midst of many obstructions. If this wire gets cut or burnt by the hot headers, the electrical system is dead. The answer is to use a Ford remote starter solenoid. The hot lead is then run from the battery positive side to the remote solenoid, which is usually located somewhere inside the race car, probably on the ignition components module panel. Then the hot lead is routed from the remote solenoid to the Chevy solenoid. This requires a little rewiring at the starter. Looking straight on to the end of the Chevy solenoid, from the front, it has four poles. The top left one (marked R) won't be used. The big one in the center receives the hot lead from the Ford remote solenoid. In addition, a copper or aluminum strip is run from this pole to the top right one (see photo). The bottom pole connects to the starter.

Above, arrow points to the aluminum strip which runs between the two top poles on a Chevy starter when the Chevy solenoid is bypassed. Right, even though the starter support strap can get in the way of headers, it should be used to prevent that heavy starter from deflecting the block casting.

THE ALTERNATOR

When choosing an alternator to be used on your engine, always be sure to get a voltage regulator that matches the alternator. If you are not buying them as a matched set, you can look on the alternator to find a plate or a stamping on the body of it which tells you the volt and amp output of the alternator. It might be 15 volts and 55 amps, or 12 volts and 40 amps, or anything else. Purchase a new regulator for those specifications, made by the same maker as the alternator (Delco Remy for Delco Remy, Motorcraft for Motorcraft, etc.)

The alternator bracket is a highly stressed piece, and if it fails, you are out of a race. Make sure it is a sturdy piece like this.

If you use one of the new alternators with a built-in regulator, you have it easier in the wiring department. In effect, the factory has done the wiring from the alternator to the regulator for you. So from there, proceed with your wiring just like it is coming out of the old regulator box.

If you ever have to fit an alternator on an engine with extreme space limitations, remember that the alternator can be placed on either side of the engine, or it can be turned around and run in the opposite direction with no problems.

Alternators are a horsepower drag **only when** they are charging. So, if you want a little extra horsepower boost when you are qualifying, splice a toggle switch into the wire connected to the field pole (terminal marked "F") on the alternator.

Use a large diameter pulley on the alternator so it does not turn at much higher RPM's than it was designed to operate at.

For short track racing, the alternator can be completely eliminated. Be sure the battery is sufficiently charged to supply the needs of the ignition and starting systems.

THE BATTERY

We'll dispense with information about how the battery works, and get right to the choosing and use of the battery.

Hot starting of a racing engine takes a huge drain on a battery, so it only makes sense to have the best available when you really need it. When purchasing a battery, you should look at the amp-hour rating of it. This indicates the electrical reserve power that it has and its power output. For a race car, you will want at least a 70 amp-hour battery.

To be sure you'll always have power available when you need it, some routine care of the battery is in order. Keep the

battery well charged and properly filled with distilled water. To test the charged condition of a battery, the electrolyte in each cell should be measured with a hydrometer. A fully charged battery will read between 1.260 and 1.280 on the hydrometer, which means the electrolyte is between 1.260 and 1.280 times as heavy as pure water at the same temperature (all readings are corrected to 80 degrees F. for electrolyte). For every 10 degrees above 80 degrees, add four points (.004) to the specific gravity reading. For every 10 degrees below 80, subtract four points from the reading. All battery cells should produce an equal reading. If the

One strap holds this battery in place, but it is a sturdy one. There should be a vertical hold-down strap too. Note grommet around positive cable entering firewall.

A well-constructed, lightweight battery box will positively locate the battery no matter what happens.

DO NOT SUCK IN TOO MUCH ELECTROLYTE

HOLD TUBE VERTICAL

FLOAT MUST BE FREE

1.270

TAKE READING AT EYE LEVEL

The procedure for checking the specific gravity of a battery with a hydrometer.

reading between any two cells of a fully charged battery varies by more than .050, the battery is bad and should be replaced.

Battery cables carry extremely heavy current during starting, so to prevent any voltage losses, as large a battery cable as possible should be used. Use no. 0 or no. 1 gauge copper arc welding cable to connect the positive battery post to the solenoid and the solenoid to the starter, to prevent current drop. Connect the negative post to the frame rail—only a short run of no. 2 gauge copper cable is required. Don't install your battery cables taut. A jolt to the chassis or a movement of the battery could mean the cables pulling on the battery. This can pull a cable connection loose, break a terminal or damage the battery. Give the cables some slack.

With the battery located in a compartment tucked away from sight, it will be easy to neglect it for regular maintenance. But even still, sulfuric acid fumes are escaping through the battery caps and attacking the cable terminals and posts, creating corrosion. This corrosion destroys the bond between the clamp and post, which impedes current flow. So keep the terminals brushed and coated with a light coating of grease.

Beware of lightweight battery boxes and supports. The heavy mass of the battery can fatigue and fail brackets, especially on dirt tracks.

To prevent turning the alternator at high RPM's a larger than stock pulley can be mounted on it.

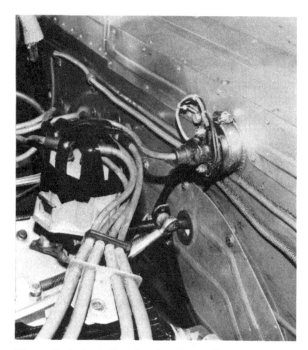

The coil is installed in the firewall to protect it from the heat of the engine compartment. It was mounted as close to the distributor as possible to prevent a long run of cable. Note the positive hold-down of plug wires to the distributor cap. The wires have been known to come loose during a race.

Hood clearance and distributor clearance conflicted here to get the proper size air cleaner mounted. The solution? Mount the air cleaner off-center from the carburetor. Works neat.

The Fuel System

THE FUEL CELL

The fuel cell is one of the greatest safety devices ever invented for automobile racing. Whether your racing association requires it nor not, a fuel cell should be used in every racing vehicle. It adds to the cost of a car, but just remember that it is very cheap insurance in terms of saving a driver from the misery and suffering of burns.

PICKING THE RIGHT FUEL CELL

There are several grades and sizes of fuel cells to choose from, so your cash outlay can meet your style of racing. Fuel cell capacities are generally available in 8, 12, 15, 22 and 32 gallon sizes. Only choose a capacity as large as you are ever going to use. However, if you choose an 8-gallon cell for a sportsman car because you never use more than eight gallons for a 35-lap main event and 10-lap heat race, just remember that you will have to refuel twice during the season-ending 100-mile race while your competitors with the 22-gallon cell won't have to worry about refueling. Generally speaking, a sportsman or modified car will usually use a 15-gallon or 22-gallon cell.

You should be aware that some fuels and fuel additives cause foam deterioration. If you plan to run something other than pump gasoline, aviation gasoline or racing gasoline, check with the manufacturer to see if your fuel is safe to use in his cell.

If you race in a division where quick refueling during pit stops is important (like Grand National racing), then a cell with exterior refueling capability is important. If you are just going to run short track events, the "slow fill" cell which refuels through a short filler neck in the trunk will get the job done just fine and save you some money.

Some fuel cell manufacturers will sell you the fuel cell system without the sheetmetal cannister which encloses it (such as the ATL Saver Cell), allowing you to make the cannister yourself and save a few dollars. If you make your own cell cannister, be sure to use at least 20 gauge steel sheet material. The fuel cell should be secured in the car with two **steel** straps (1/8-inch thick, 2-inches wide) widthwise and two straps lengthwise, both top and bottom if the cell is not held in a belly pan.

NASCAR and some other sanctioning bodies requires that a 20-gauge steel sheetmetal belly pan be placed under the fuel cell cannister if the cell hangs below the top of the frame rails. In this case the steel straps are only required on top of the cell.

LOCATING THE FUEL CELL

Because the weight of the fuel supply in the fuel cell decreases as gasoline is consumed, the fuel cell should be located as far foward as possible in the car (toward the center of gravity) to minimize the effect of the change in weight distribution. This is also a good idea safety-wise so the car has a margin of crushable structure before the rear bumper contacts the fuel cell, in case of an accident.

The fuel cell should be located in the car as low as

Principle of Operation

A "Fuel Cell System" is a safety fuel container which replaces a conventional gasoline tank. Fuel cells suppress the leakage, fire, and explosion that otherwise would occur when a vehicle crashes and its tank ruptures. In race cars, approximately 95% of the potential fuel fires have been eliminated through the use of these safety cells.

The Fuel Cell System is comprised of four primary elements:

1. A tough, but flexible, rubberized fabric bladder with outstanding tear and puncture resistance.
2. Anti-explosion foam baffling throughout the bladder interior.
3. A metal container surrounding the bladder for flame resistance and ease of installation.
4. One-way check valves which seal off the filler and vent in case of overturns.

The accompanying diagrams depict how a fuel cell system works in an accident situation.

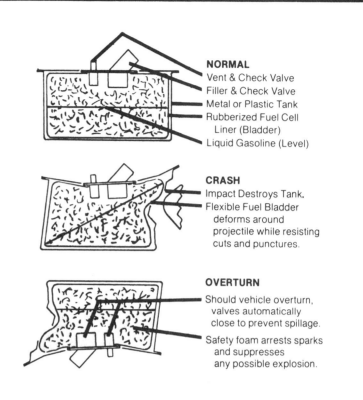

NORMAL
Vent & Check Valve
Filler & Check Valve
Metal or Plastic Tank
Rubberized Fuel Cell
  Liner (Bladder)
Liquid Gasoline (Level)

CRASH
Impact Destroys Tank.
Flexible Fuel Bladder
deforms around
projectile while resisting
cuts and punctures.

OVERTURN
Should vehicle overturn,
valves automatically
close to prevent spillage.

Safety foam arrests sparks
and suppresses
any possible explosion.

The photos above and to the right illustrate how 2-inch by 1/8-inch steel straps should be used to securely anchor the fuel cell. The straps attach to an angle iron frame which is welded between the frame rails. The underside of the cell here is protected by a sheetmetal belly pan.

possible so the fuel load will not have a detrimental effect on the center of gravity heighth, up to a point where a low-hanging cell is in danger of being hit or knicked by debris.

When mounting the fuel cell bladder in the cannister, be sure to place the fuel pick-ups pointing toward the outside rear of the vehicle so there is always a positive fuel feed during acceleration.

THE FUEL FILTER

Gasoline should be filtered just as soon as it is pumped from the fuel cell. Use a Fram HPG-1 gasoline filter, a large capacity, low restriction, remote-mounted filter, in the fuel line, mounted in the trunk area. This is a replaceable cartridge-type filter. This is especially important if an electric fuel pump is being used (filter before the fuel reaches the pump). This removes fuel impurities before they enter the line or pump, and also saves a lot of fuel line plumbing problems by eliminating in-line fuel filters by the carburetor. In-line filters have a large restriction rate, and can be easily punctured.

The Fram HPG-1 fuel filter mounted in the trunk area to filter the gasoline just after it is pumped out of the fuel cell.

THE FUEL LINE

The fuel line should have a minimum inside diamter of 3/8-inch. A half-inch inside diamter is even better. If a steel braided neoprene line is used, choose the -8 size.

The steel braided neoprene line is the ultimte choice for use as a fuel line. It has a built-in armor-plated exterior, which also helps prevent cracking due to vibrations and flexing. A regular automotive neoprene fuel line can be used

from the fuel cell to the fuel pump on the engine, but it should not be routed underneath the car where rocks, trash or scrapes from off-track excursions could damage it. Route it through a pipe passing through the interior of the car (use thin wall electrical conduit). From the fuel pump to the carburetor, the use of steel braided line is highly recommended. The routing of the fuel line should be planned with a minimum of kinks and bends involved.

Hose clamps are fine for passenger car use, but in a race car fuel system, only threaded connections should be used. Impacts can pull a clamped line apart.

ELECTRIC FUEL PUMPS

Most racing associations look upon electric fuel pumps as a fire hazard, and as such are banned from many race cars. If you are allowed to use them, however, place the fuel pump as near to the fuel cell outlet as possible. The electric fuel pump is much more efficient in pushing fuel, rather than pulling it. Because of this, electric fuel pumps are good at preventing vapor lock in the lines by keeping pressure in them.

Use a roll-over mercury switch in the electric fuel pump for roll over protection. All mercury switches are not necessarily roll over switches, so specify.

FUEL PRESSURE GAUGE

Constant fuel pressure to the engine is an important item to monitor, so a race car should be equipped with a fuel pressure gauge. The gauge is especially important during initial testing of a race car. If the fuel pressure fluctuates, it could indicate a problem such as a kinked line, pick-up problems in the fuel cell, or a clogged fuel filter. The fuel pressure should be measured at the fuel inlet to the carburetor. Minimum fuel pressure should be 4 PSI at high RPM, wide open throttle.

It is strongly recommended that the fuel pressure line to the gauge be an Aeroquip steel braided line with a thread coupler for driver safety. Remember that the fuel pressure line is an unprotected gasoline line in the driver's compartment. Care should be used not to route the hose over electrical wires or sharp objects.

GASOLINE FOR RACING

Service station premium gasoline, for the most part, is inadequate as a racing fuel for engines having a compression ratio of more than 11 to 1. The pump gasoline available today is a bad product in terms of octane quality required for racing engine, being filled with unwanted byproducts such as butanes and very heavy hydrocarbons. The butanes are very light bodied substances which promote easy starting of a passenger car on a very cold morning, but they also promote vapor lock in a race car. The heavy hydrocarbons exist on the opposite end of the scale from butanes, and contribute to deposits in the engine which can lead to detonation.

Another method of securing a fuel cell is illustrated here.

There are "junk" products found in some pump gasolines, hidden in the fuel by some refiners to dispose of excess quantities on hand. These products, such as naphthene, asphalt and solvents, are mixed into gasoline in a solution of roughly two percent by volume. You will generally find these types of gasolines at independent gasoline stations, where the fuel cost is less per gallon, and where no quality control is assured to the buying public.

Adding to the problems of pump gasoline is the fact that it is almost impossible to get the same tank full of gasoline at the same service station twice. When you travel a distance to a race, the problem compounds itself. The refineries are always changing blends of gasoline, to suit weather and seasonal changes. Gasoline blends also change by elevation and region in which they are marketed.

The blends mostly vary by their volatility. Octane quality, when purchased from a major refiner, usually remains constant. Volatility is the ease of the fuel to turn into a vapor. The lighter bodied the gasoline is, the easier it vaporizes. The problem with volatility in a race car is that the gasoline could very well vaporize in the lines before it reaches the carburetor. The fuel pump cannot pump vapors, so you have vapor lock. Then gasoline cannot be pumped to the carburetor. Volatility is controlled by additives such as butane, propane and methane. When checked with a gasoline hydrometer, very volatile gasolines will have a smaller specific gravity.

AVIATION GASOLINE

To solve the problems created by the deplorable octane quality of pump gasoline, many racers have turned to using pure aviation gasoline, or a mixture of aviation and premium pump gasolines. This is a partial, but not a complete answer, and without some guidance, these people can damage a racing engine with aviation gasoline.

Aviation gasoline can be nearly a 100 percent pure hydrocarbon (the more desirable part of the crude oil refined for use as fuel).

It will produce as many BTU's of heat as a pump gasoline, but because aviation gasoline is less dense, it will not produce as much energy as pump gasoline when burned. So to be on an equal footing with pump gasoline, more aviation gasoline is needed in the combustion chamber to do the same job. This means the carburetor must be re-jetted one step richer to accommodate pure aviation gasoline, or else the engine will run lean and burn pistons. Fuel injected engines will have to be richened a couple of steps. It is best to use aviation gasoline blended with heavier-bodied hydrocarbons to increase the specific gravity of the fuel.

There are four grades of aviation gasoline which you should be aware of. The most common is designated 100-130. Because aviation gasolines are knock rated under differing standards than those used for automotive gasolines, the 100 does not mean 100 Motor Octane Number, and the 130 does not mean 130 Research Octane Number. The 100-130 aviation gasoline has a 100 MON and a 103 RON. This grade of gasoline is the one generally desired for use in a racing engine.

Another grade of aviation gasoline is designated 80-87, and it definitely is **not** to be used in a racing engine. It is an unleaded low octane aviation gasoline, which has octane ratings lower than those for automotive premium pump gasoline.

There is an aviation gasoline designated 100 Low Lead, and it is virtually the same fuel as 100-130 av gas. The only difference is that is has less lead in it, but there is still enough in it to protect valves in a racing engine. The 100 Low Lead is perfectly acceptable as a racing product.

The fourth designation of racing gasoline is 115-145. This is a very high octane, very high lead content, military gasoline. It is not generally available to the public. If you can get it, however, it is a good racing fuel.

You may find a difficulty in purchasing any aviation gasoline at an airport without providing the dealer with an airplane number. This is because the dealers fear that people are using the av gas in automobiles on the road, and this presents a legal entanglement as any fuel which is used in a road vehicle must have road tax paid on it. There are no road taxes attached to av gas. If you should find yourself in a predicament, convince the fuel dealer that you are purchasing the gasoline for use in a race car which will not be used on the road. To make' the dealer's time more worthwhile, too, be sure to purchase a 55-gallon drum full of gasoline at one time. A five-gallon sale for him is not worth the trouble.

RACING GASOLINE BLENDS

Racing gasolines, blended specifically for high RPM racing engines, start with a known, controlled quantity like the aviation gasoline, then heavier-bodied hydrocarbons are added. These hydrocarbons include elements like isopentane, isooctane, benzene, toluene and isohexane, along with a greater amount of tetraethyl lead. They are blended in a specific formula which never varies, giving the racer a known quantity with each purchase.

One of the premier makers of racing gasoline is Union Oil. They supply racing gasoline for all NASCAR Grand National stock car races each year, plus a number of the major racing events. They sell this same blend of racing gasoline in 55-gallon drums through selected distributors troughout the nation.

There are also several companies which purchase stocks of aviation gasoline and add their own formula of heavier-bodied hydrocarbons. Some of these include Daeco Racing Gasoline in Wilmington, Calif., Golden West Racing Fuel in Sherman Oaks, Calif., H & H Racing Gasoline in Terre Haute, Ind., So-Cal Racing Fuel in Torrance, Calif., and Volitile Products Co. in San Antonio, Tex.

For any racing gasoline, be prepared to pay at least one third to one half more per gallon than the going rate for premium pump gasoline.

AN ALTERNATIVE GASOLINE

An alternative to racing gasoline is to blend your own. The recipe starts with a clean 55-gallon drum (we strongly discourage the use of a bath tub!). Add in it 25 gallons of the finest premium pump gasoline you can find, 25 gallons of 100-130 aviation gasoline and 5 gallons of toluene. The shelf life of this blend is about one year. Be sure the drum is kept as full as possible, and that it is very tightly sealed. Contamination with air and water will spoil the blend.

The aviation gasoline blended with the pump gasoline assures a purer hydrocarbon content, and assures a lower vapor pressure (less volatility). The toluene, a very heavy-bodied substance, acts as an octane improver (it has 103 MON octane and 120 RON octane). The end result mixture will give you high octane, low volatility, and a specific gravity the same as pump gasoline so you will not have to rejet the carburetor. The most important thing is that you have a standard blend which will not vary.

Toluene is generally available at paint stores, or chemical supply houses. If you should have any indications that detonation is occurring in your engine, add a little more toluene. When reading spark plugs and looking at the exhaust ports, expect to see a soft gray sooty appearance caused by the toluene. At first you might mistake it for an over-rich condition. Be very careful when handling toluene—it is a very toxic material and can be absorbed into the body through the skin.

OCTANE IMPROVERS

There are many octane improvers on the market which you can pour into the gasoline. These generally are based on aniline, which is a compound very similar to toluene and benzene. It improves octane quality, and does not add to fuel energy or power. It also is very toxic and expensive. It can be added to the racing gasoline or alternative blend gasoline if you wish, but it probably will not do any good. This is because its only function is to improve octane quality, and once the octane of a fuel is improved to a point where knock disappears, its effectiveness diminishes and no more power can be gained. Adding any more octane improvers to the fuel past this point will only serve to increase the build-up of deposits in the engine, which can

Properties of various fuels and additives used for racing.

Fuel	RON	MON	Lead grams/gal	Specific Gravity	RVP	Availability
Premium Gasoline	97-99	90-92	1.5-3.0	0.73-0.76	7.0-12.0	Gen. Available
Racing Gasoline	104-105	97-99	3.8-4.2	0.72-0.75	6.5-7.0	At Tracks
Aviation Gasoline 80/87	85-87	82-84	0.4-0.5	0.70-0.71	5.5-7.0	Airports
100/130	103	100	3.4-3.7	0.69-0.71	5.5-7.0	Airports
115/145	107	105	4.4-4.6	0.69-0.71	5.5-7.0	Not generally Available
Methanol (Alky)	107-112	87-91	-------	0.79	4.5	Chemical Supply House
Toluene	116-120	100-103	-------	0.87	1.0	Chemical Supply House
Benzene	106-110	95-100	-------	0.88	3.2	Chemical Supply House
Nitro-Methane	<0	<0	-------	1.14	<1.0	Racing Fuel Outlets

eventually lead to detonation.

Do not purchase methanol to blend with gasoline as an octane improver. It can separate from gasoline if any hint of water is present. The addition of methanol also requires significant jetting changes.

Nitromethane is another compound which is not desirable to be mixed with gasoline. It lowers the octane number of pump gasoline.

Racing gasoline should be stored in a tightly closed, clean container. Be sure to plainly label the contents of the drum so nobody accidentally contaminates the gasoline.

The Cooling System

At the heart of the cooling system is the radiator, and for any race car, it must be custom-tailored to the job. This does not mean that every race car must have an expensive custom-made radiator, but rather that the capacity and efficiency of the radiator must be specifically chosen for its application.

There are several factors which determine the efficiency of a cooling system. The most important is the coolant capacity of it. Some people may get carried away with this fact, thinking if some cooling capacity is good, a lot of capacity is great. But, a radiator capacity is sufficient if the coolant flows through the radiator slow enough to allow effective transfer of heat through the tubes and fins.

A second determining factor in the efficiency of the radiator is its physical dimensions. A narrow radiator with six rows of tubes is not going to be efficient. The air passing through the back three or four rows of tubes will be carrying a lot of heat which has already been transferred to the

This custom-made radiator [from a local radiator shop] features a wide frontal area, 3-tube depth and crossflow design. Notice how the radiator is mounted at the top in a single, sturdy, rubber-cushioned bracket.

airstream from the front rows of tubes. Only the front portion of this radiator would be doing an effective job, so the radiator would be holding a capacity twice its efficiency rate.

Another factor effecting the efficiency of a radiator is its frontal area. This means the radiator should be as tall and wide as possible to give it sufficient and unobstructed exposure to cool air.

As the horsepower output of an engine increases, so must the efficiency of a cooling system. Extra horsepower means extra heat which must be carried away by the cooling system. Many racers find that a sudden increase in engine horsepower is usually greeted with overheating problems.

UPRIGHT VERSUS CROSS FLOW

For racing applications, the crossflow radiator represents a much more efficient system for cooling. Its greatest advantage is that the coolant is forced to flow **across** the radiator — gravity is not pushing it. Flow from one side to the other is much slower than flow would be from top to bottom on an upright radiator. The ultimate effect of the slower flow is that the water has a longer time to dissipate heat.

THE CORE

There are several styles of radiators, but the most common to auto racing vehicles is the tube and fin design. In the tube and fin type of core, the tubes are oblong in cross-sectional shape which gives the coolant as much exposure to the wall area as possible. The fins are important in the heat transfer process in that they conduct the heat away from the tubes. Radiator cores are available in fin counts (meaning the number of fins per inch) ranging from 7 to 20 fins per inch. The greater the fin count, the better the heat dissipation capacity of the radiator (as long as air flow through the radiator is not restricted).

Although heavier duty truck parts are generally accepted as being better designed for racing service, such is not the case with heavy duty truck radiator cores. Most truck cores have thicker-wall materials in the tubes to protect against the greater abuse and vibration a truck has to withstand. So, only passenger car cores should be used to insure proper heat dissipation.

In considering any core for use in a race car, three to four rows of tubes should be deemed sufficient thickness for proper cooling. The more important factors would be larger surface area (greater width and heighth) and the fin count.

There are two approaches you can take when purchasing a radiator for a race car. The first is to find a passenger car radiator from an existing car which has sufficient size, capacity and efficiency and which has dimensions applicable to your vehicle. Some possibilities of these would be a four-tube crossflow from a Buick Wildcat, a Jaguar XKE aluminum crossflow, or a four-tube crossflow from a Pontiac station wagon with air conditioning. And of course, there is the good old reliable racer's standby, the Corvette

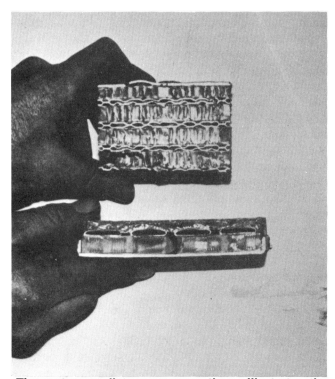

These two radiator cross sections illustrate the difference in tube construction. The top section is from a Corvette and utilizes stamped sections which are furnace brazed. The bottom section is a Holman and Moody racing radiator — big tubes, huh?

aluminum crossflow. When purchasing a Corvette radiator, know which one you are getting. There are seven different sizes available. The one most desirable is the one which was used with the 427 and 454 cubic inch engines.

The second approach is to have a radiator repair shop build one for you. What this entails is having the shop order a core for you, then they solder the tanks onto it. Most radiator shops have access to several core manufacturers who can supply any type and size core needed, plus they can custom-make anything special you may need. Once the tanks and core have been soldered together, the radiator should be pressure-checked.

For further care of the radiator to keep the core tubes clean and open, have it boiled out and pressure checked at least once during the racing season (in the middle of the season), and then again during the off-season.

A tip about soldering: use a solder which is 60 percent tin, 40 percent lead. If the lead content is any higher, a tank or fill fitting can crack and blow right off at the solder joint when subjected to high pressure and temperature.

KEEPING IT COOLER

There are four basic ways to tune the cooling system to get the ultimate in cooling efficiency from it. They are: 1) to increase the pressure capacity of the system 2) maximize the ducting and shrouding, 3) vary the selection of the fan size and pitch, and 4) adding varying amounts of a coolant to the water.

PRESSURE CAPACITY

The normal boiling point of the coolant contained in a cooling system can be raised without the coolant boiling by increasing the pressure capacity of the system. At sea level the atmospheric pressure is 15 pounds per square inch, and water boils at 212 degrees F. Each pound of increased pressure over atmospheric pressure contained in the cooling system raises the boiling point of the coolant by 3-1/2 degrees F., thus enabling the coolant to circulate in the system at a higher temperature without boiling. The operating temperature of the engine is relatively unimportant, as long as the coolant does not boil. Once it does, coolant is pushed out of the system, air pockets are formed, and heads will then crack.

The pressure is maintained in the cooling system by the radiator cap. The cap is equipped with a spring whose rate is calibrated to relieve the pressure build-up at a predetermined value. If you plan to use a high pressure cap (between 17 and 21 PSI is recommended), be sure that you have a high quality radiator which will contain the high pressure, high temperature and vibrations. It is a good idea to have a radiator shop pressure check the radiator at the operating pressure you intend to use.

SHROUDS AND DUCTING

Proper airflow through the radiator is vitally important to get the most out of a radiator's efficiency. When the airflow is left to chance without the direction of a shroud and ducting, the air can actually flow around the radiator, or through just a small section of the radiator, hindering the total heat exchange capacity of it greatly. To achieve a proper airflow through a radiator, the air must be ducted into it, and a fan shroud must be used behind the radiator. A shroud, which is designed to taper down from a large opening to a smaller one, will force the air to pull from the entire frontal surface area of the radiator. A side benefit of a shroud is that the air is cooled slightly as it is forced down into a smaller area when being pulled through the shroud. Also, it allows the fan to be placed further from the radiator.

In front of the radiator, the air must be ducted to it from the grille to prevent the airstream from being deflected and spilling over the top or around the sides of the radiator. Use .040-inch thick aluminum sheet panels to build the air ducting, and be sure to seal all the seams of the ductwork, as well as where it contacts the radiator, with weather-stripping.

For cars using a large front spoiler under the grille or bumper, it would be advantageous to duct air from the top of the spoiler into the radiator. This is especially an advantage on dirt tracks where mud can clog the front grille opening.

On superspeedway cars, the ducting from the top of the spoiler into the radiator is the major source of airflow for the radiator. The front grille opening on superspeedways cars is generally sealed almost completely with several layers of screen.

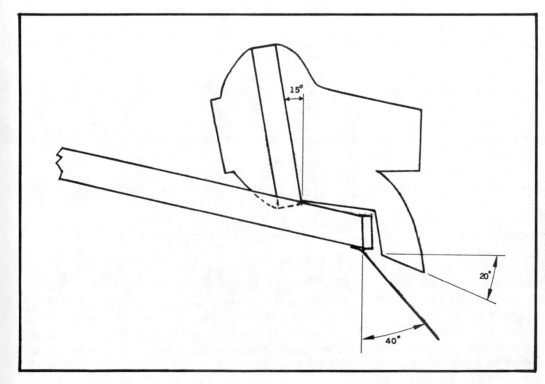

The drawing illustrates the proper method of mounting the radiator [15 degrees laid back] with a fan shroud behind it, and ducting which directs the air into it from the grille and the front spoiler.

This huge radiator is required to cool a 454-cubic inch big block. Note the ducting leading into it. Also note that the oil radiator is mounted so it is in the air stream, but separate from the water radiator.

THE FAN

The fan's job is to **pull** fresh, cool air through the radiator core, with the air carrying away engine heat being conducted to the tubes and finds from coolant. The determining factors of fan efficiency are: 1) the fan blade circle diameter, 2) the blade pitch, 3) the number of fan blades, and 4) the proximity of the fan to the radiator.

Ideally, the fan should be located one inch from the radiator core if no fan shroud is used. And, this means one inch over the entire fan diameter, not just at the top. Any further away than this and the fan will suck air from around the sides of the radiator rather than through it. If a shroud is used, the fan can be placed at the back opening of the shroud. You'll be surprised how many radiators this can help save over a racing seson.

Blade pitch and fan diameter are very closely related. A small diameter fan needs to have more pitch to draw more air. The greater diameter fan needs less blade pitch to draw the same volume of air. The greater the pitch of the blades, the more air the fan draws, but the more horsepower required to turn the fan.

A good fan specification to use for a starting point for a short track racing motor is a four-blade fan, with 15-1/2-inch blade circle diameter and moderate blade pitch. Depending on the type of track conditions you encounter, you may need to alter this. For example, if you run on a small tight track where one car has to follow another for a number of laps before passing (meaning a blocked airflow through the grille for a number of laps), then more blade pitch or diameter may be required. On a dirt track where the grille plugs up with mud, at least a five blade fan will probably be needed. On the other hand, if your engine runs at an acceptable operating temperature but heats up only while idling, find a fan with just a little more blade pitch to solve the problem.

On superspeedways, virtually no fan is required. In fact, if it were not for pit stops, superspeedway cars would not need a fan. There are other places where this can also be true, such as a road course or a fast 1/2-mile or 5/8-mile track.

Even with a fan spacer and large pitch fan, there is still quite a distance from the fan to the radiator. A shroud will help greatly.

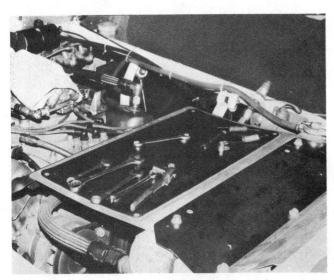

A handy use for the top of the fan shroud is as a tool tray.

ANTI-FREEZE CHART

Anti-freeze solution by volume (%)	Boiling point with 18 # pressure cap
30	268° F.
40	270° F.
50	273° F.
60	278° F.
70	284° F.

In searching for the right fan, you can consider the entire stock of GM and Ford fans if you have one of these engines. The four-bolt fan pattern on both Ford and GM engines is an interchangeable pattern.

Fan belt tension is very critical with large pitch fans, because loads are very high on the belts. And for this reason, do not consider using a seven blade, high pitch fan. This type of fan was designed to run on a luxury automobile at 2300 to 2600 RPMs. Over 3000 RPMs it begins to stall, and at 7000 RPMs the engine is just slipping the fan belts.

COOLANT

At least fifty percent of the coolant solution in the cooling system should be a permanent-type ethylene glycol base anti-freeze solution. The addition of this anti-freeze solution significantly raises the boiling point of water (see chart). It also prevents water from foaming (foaming means steam and air pockets in the system), prevents corrosion in the engine, and helps to lubricate the water pump thrust seal. If you switch from an all-water coolant to a fifty percent glycol coolant, you will notice that the engine will run about ten degrees warmer. This is because glycol retains heat. But it also raises the boiling point, so the increased operating temperature is insignificant.

Do not add any stop-leak compounds to the coolant. If you have a leak, have a radiator repair shop fix it the right way. The stop-leak compounds will gum up the tubes and reduce the cooling efficiency of them.

MOUNTING THE RADIATOR

The radiator, in spite of all its bulk and metal content, is still a delicate piece of equipment and should be treated as such. In planning its attachment to the chassis keep in mind that the soldered joints of the radiator can fatigue quickly if it is not mounted correctly. At the bottom, the radiator should be mounted in a cradle made of steel strap which is then cushioned with rubber insulating material. It should never be solidly mounted. Likewise at the top, rubber insulated mounts should hold it in place and protect it from vibrations and flexing.

The radiator is always quite vulnerable to punctures from debris, so it should be adequately shielded in front with wire mesh screen.

To improve cooling efficiency, the radiator mounts should be designed to allow the radiator to lay back at the top 15 degrees from vertical. It improves the angle of attack of the incoming airflow.

HOSES

Good quality radiator hoses are very important to protect the investment you have in your racing engine. If rubber hoses are used, they should be the smooth-wall type and not the ribbed-wall design, to protect against flow restriction, and the hoses should have a spiral wound wire in them to prevent hose collapse at high RPM.

In recent years, the quality of automotive radiator hoses has fallen. They cannot be trusted past 12 months, and on a race car, should be replaced at least once during a racing season. In a race car application, the hoses are facing much higher pressures and temperatures than they would in a passenger car. Be sure to constantly look for signs of trouble, such as a bulge in the hoses.

Automotive rubber hoses should be taped while in installed position with nylon fiber-reinforced strapping tape, then a layer of black electrical tape. This reinforces the wall of the hose, helps the hose retain its shape (as in curved hoses), and protects the hose from punctures caused by small rocks or other debris.

Rather than adapting a passenger car radiator hose to a race car, try these alternatives. First, truck radiator hoses are usually of higher quality. They could get the job done for you. Second, try one of the industrial-use silicone rubber-based hoses. These are made by Flexfab Inc. in Hastings, Michigan, and by Stratoflex in Fort Worth, Texas. Finally, the ultimate (but most expensive) answer is to use steel braided hose (-16 size).

Use only worm gear-type clamps with the slots punched

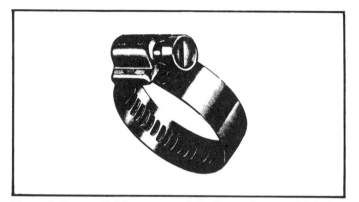

Use only tangential-type worm drive hose clamps with holes punched all the way through. This one only has drive clips on the inside of the band — no good.

all the way through the clamp to secure radiator hoses, and use two clamps on each end of hose. As an added safety margin, tighten the clamps while the engine is hot. To make the hose installation and removal easier, coat the radiator neck and thermostat neck with a silicone-based grease.

Make sure the lower radiator hose is not the lowest part of the chassis, or you will find yourself in hot water when running over debris, dirt clods and rocks, or off the track. Also be sure to weekly check the lower hose for signs of debris damage, even if the lower hose is up out of the way.

THE COOLANT RECOVERY TANK

When an engine overheats, boils the coolant and forces it out of the system, air and steam pockets are formed, and this can crack cylinder heads. A simple preventive measure for this problem is to have a completely closed cooling system. This employs the use of a coolant recovery tank.

The recovery tank catches and holds the hot expanded coolant which is pushed out of the system, then allows the cooled fluid to be siphoned back into the radiator once the engine cools.

Another benefit of the tank is that it keeps coolant from being dumped onto the racing surface, or under the wheels of the race car.

The simplest and cheapest form of recovery tank is the small plastic variety designed for passenger cars, available for a couple of dollars at most any auto parts store. Be sure to solidly reinforce the flimsy plastic tab mount most of these tanks have. Another popular tank in use on race cars is the surge tank used with the Corvette aluminum crossflow radiator.

When the radiator is lower in the chassis than the top of the motor — as in Corvettes and the Plymouth Super Birds — a surge tank is required. The surge tank is mounted as the highest point in the cooling system, and the system is then filled from the surge tank.

A surge tank may be used even if the motor is lower than the top of the radiator. It offers the advantage of about one additional pint of coolant in the system, plus a remote fill (if

the front of the car has aerodynamic sheetmetal preventing easy access to the top of the radiator), and eliminates the possibility of air lock in the engine if the system is filled in a hurrry.

The coolant recovery system may be used with either a conventional cooling system or surge tank system.

The Harrison/Corvette aluminum surge tank is mounted in a convenient place, and as the highest point in the cooling system. The ribbed wall top radiator hose hinders coolant flow. Note tape wrapping on wires from alternator — a good protection against heat.

PULLEYS AND BELTS

In a racing application, most pulleys and brackets for driving accessories such as the water pump, alternator and possibly a dry sump pump and fuel injector pump, will be changed and relocated from the stock position. When this happens, there is a chance of getting the pulleys out of alignment. Once all the pulleys and belts have been installed, sight along their running plane from the side and top to be sure they are all in alignment. If they are not, use spacers to align them, or you will be experiencing thrown belts or even a sheared water pump shaft.

The pulley attached to the water pump on the racing engine will have to be changed to modify the water flow speed. On passenger cars, the water pump is turned from 1.5 to 2 times as fast as the crankshaft speed, to promote cool operating temperatures while the car is idling in traffic.

If the water pump were to turn faster than the crankshaft speed, the coolant would be flowing far too fast to pick up enough heat in the block and to radiate the heat in the radiator. The result would be overheating.

Because pulley belts tend to slip when turned at high RPMs, an alternative idea to the conventional pulley system is the use of cogged gear pulleys. Another advantage of the cogged drive pulleys is that they can save a substantial amount of weight. Aluminum cogged gear pulleys (and belts for them) are available from Weaver Brothers in Van Nuys, California. A source for steel cogged gear pulleys is from Vega and Pinto engines. Gapp and Roush in Livonia, Michigan, make aluminum pulleys in various sizes to fit the Pinto engines. The cogged pulleys from different manufacturers are not interchangeable, however, because the teeth spacing on the pulleys is different.

Complete cogged gear pulley systems like this are available from Weaver Brothers.

A cogged gear pulley attached to a Chevy small block water pump.

If automotive V-belt pulleys are used instead of the cogged gear drive pulleys, the only brand name belts to use for racing is Dayco. Several years ago Dayco introduced a special racing design V-belt which was given to and used by all the Grand National teams. Dayco engineers wanted to try and prove several design techniques as well as materials with this promotional experience. The belts proved themselves in tough competition, so Dayco decided to incorporate this special racing design into their full line of automotive V-belts. So, any time you purchase a V-belt at the auto parts store with the Dayco name on it, you are getting the special quality belt equivalent to Dayco's previous high performance-designated belts. These belts incorporate higher quality synthetic fabrics in the tension area, have a high tensile strength, have a greater compression area (which means less belt slip at high RPM's) and have much less tendency to flip over at high RPM's.

Don't settle for a V-belt with another name on it which a parts counterman says is made by Dayco. It may be, but it is only the top quality belt when it has the Dayco name on it.

THE THERMOSTAT

The major purpose of the thermostat is to control the minimum temperature experienced in the engine. When the engine and coolant are cold, the thermostat stays closed and coolant is circulated only through the block and not the radiator. When the coolant reaches a temperature regulated by the preset rating of the thermostat, the thermostat opens and the coolant is circulated through the radiator.

Another purpose of the thermostat is to act as a restrictor to regulate the flow rate of the coolant through the block. If coolant flows too fast, it will not have a chance to pick up all the heat it should which is being transferred from the block.

Most top engine builders feel that the thermostat regulating mechanism should be removed from the housing and the engine should be operated without it. Racing engines are designed to operate at a specific temerature. The less time the engine can be operated in any other temperature range (either hotter or colder), the longer the

Always sight along the belts and pulleys to spot any misalignment. Two belts are used here to cut down on slippage.

engine is going to last. Thermostats are not reliable when subjected to racing temperatures and water flow, and can remain partially or completely closed. When the thermostat is removed from the housing, a 7/8-inch AN washer should be installed in the housing as a coolant flow restrictor. You may find that the washer opening size will have to be changed for one greater or smaller, depending on the water pump drive speed, radiator size and efficiency, and desired engine operating temperature. Quick engine warm-up can be achieved by draping a fender cover over the front of the car.

HELPING THE COOLING SYSTEM

One of the simplest ways to aid the cooling system in doing its job efficiently is to paint the engine and radiator with a light coating of flat black paint. Black colored surfaces have a high degree of thermal emission, or in other words, a flat black colored surface helps substantially in carrying away heat. White colored and polished surfaces emit heat very poorly, but reflect heat very well. So, you will want to paint the headers white to prevent heat from radiating from them into the engine compartment. And don't use a chromed oil pan, chrome rocker covers or a chromed radiator. The chrome surface will just keep the heat inside the places you want it to leave.

Oil coolers are a must for a racing engine to help the cooling system, as oil acts as a cooling fluid to assist in dissipating combustion heat. Heat which would otherwise be transferred from the oil to the block, is carried away to the oil cooler and transferred to the airstream, relieving the cooling system of a larger heat problem.

The engine coolant temperature and engine oil temperature are highly interrelated. The proper handling of one can greatly influence the other. For short track racing, here is a trick worth considering: use a stock Corvette aluminum crossflow radiator, then use a **large** engine oil cooler, mounted somewhere low in the car behind the driver. The net result is that a small light radiator is installed in front and up high, while an equal or larger portion of the cooling burden is being carried by a heavier radiator installed rearward and down low. This helps to lower the CGH and bring static weight rearward. The one trick to make this work is to be sure to have an adequate ducting of cool air through the oil radiator.

COOLING SYSTEM PROBLEMS

A problem which everyone with a race car is bound to face at least once is a leaking or blown head gasket. If symptoms lead you to suspect this problem, there is an instrument called a "combustion leak tester" which will help diagnose it. It is a syringe-type of instrument which tests the coolant for the presence of carbon monoxide in it, which of course means compression leaks. This instrument is inserted into the radiator neck while the engine is idling, and its bulb is pumped a couple of times to take a sample. If the blue solution in the instrument turns yellow, there is carbon monoxide present in the system. Most radiator shops have this instrument, or it can be purchased from Snap-On tool dealers.

Monitoring the engine operating temperature while racing the car can help spot overheating problems before they can cause engine damage. Remember that with the addition of anti-freeze and a high pressure radiator cap, the engine can operate above boiling temperature (212 F.) without serious problems. But when the engine is operating at a constant 230 degrees F., it may be advisable to shut it down. At this temperature the moly filling in moly-filled rings begins to come apart, and other problems begin to creep into the engine. And if you are watching the engine operate at 230 degrees for a while, and then the temperature gauge gets erratic or suddenly drops down to about 180 degrees, stop the engine immediately. It usually means the engine is out of coolant.

The Plumbing System

The plumbing system includes all hoses, tanks and reservoirs as well as the small connecting hardware such as clamps, tubes and fittings. A great deal of the plumbing system deals with fluid storage and transportation in the cooling system, so a great deal of plumbing knowledge is contained in that chapter elsewhere in this book. This section will discuss the choice and fitting of plumbing lines.

STEEL BRAIDED HOSE

Stainless steel braided hose is both functional and practical for plumbing fuel, oil, braking and cooling systems while adding a touch of class to the race car. Besides looking nice, it is also nearly impossible to break, wear out, overheat, come apart or assemble wrong. It is also quite lightweight.

Inside the stainless steel fiber braiding, there are two types of inner liner material: Buna-N neoprene and teflon. For most all applications, the more flexible and less expensive neoprene liner is fine. However, in extreme pressure situations like the brake system, only teflon hose liner should be used because it resists expansion **much** better than neoprene.

When working with steel braided hose, it is important to understand the nomenclature of the size designations. The hose diameter is commonly measured in "dash numbers", for example, "dash two, dash five and dash ten." The number following the dash is the number of sixteenths of an inch of hose diameter For example, dash two (or −2) is

2/16 or 1/8-inch, dash five is 5/16-inch, and dash ten is 10/16 or 5/8-inch. The diameter this represents is the hose inside diameter equal to the inside diameter of a pipe with a similar outside diameter. Got that? What this uncomprehensible explanation means is that, for example, the dash six (3/8-inch) steel braided hose has the same inside diameter as a steel pipe designated 3/8-inch outside diameter. Consult the table for a complete listing of inside and outside diameters of all dash sizes.

The best part about steel braided hose is that it makes the engine compartment look clean and professional. Aside from that, the hose adds a lot of insurance to the car.

38

HOSE ENDS

There are several brands and varieties of hose ends or threaded couplers for steel braided hose on the market. The four most popular types are distributed by Aeroquip, Edelbrock, Dave Russell's Race Car Parts and Earl's Supply. Get their catalogs to decide which fits your needs.

Threaded couplers with pipe thread should be wrapped with teflon tape around the threads for better sealing when they are being assembled. If you find it exasperating to use the teflon tape on pipe threads, use Permatex's Thread Sealant With Teflon (part #14) on the threads. It seals, lubricates, prevents corrosion and has an anti-sieze agent. If you use teflon tape, use care to prevent any tape ends or shreds from getting into the lines. It could pass on through the lines and partially clog a part.

On AN threads, use oil, anti-seize or thread lube to prevent thread galling.

PLUMBING THE BRAKING SYSTEM

Plumbing the braking system involves the use of two kinds of lines, the solid lines and the flex lines at the wheels.

The solid brake tubing lines should be Bundyweld steel tubing in the 3/16-inch o.d. size, and not copper tubing. Copper tubing will not resist corrosion in use in the braking system, and will fatigue and crack as well. The Bundyweld steel tubing is a thick wall tubing which helps resist line expansion under pressure.

The layout of the solid tubing run from the master cylinder to each wheel should be carefully designed to prevent the presence of bends and loops which will trap air. In addition, to aid the bleeding procedure, the lines should

Hose ends are available in all sizes and types to fit any need.

run consistantly downhill from the master cylinder to the breaking junction at each wheel for the flex lines. This insures there is no place present in the lines to trap air.

At connecting points of the steel Bundyweld tubing, the fitting flares should be the standard automotive double flare. The double flare was designed to prevent tubing fractures at the flare. When cutting the steel tubing in preparation for

The brake tubing should be routed so that it travels constandly downward from the master cylinder, and without any kinks or bends. This prevents trapped air. Note the supporting bracket for the tee fitting.

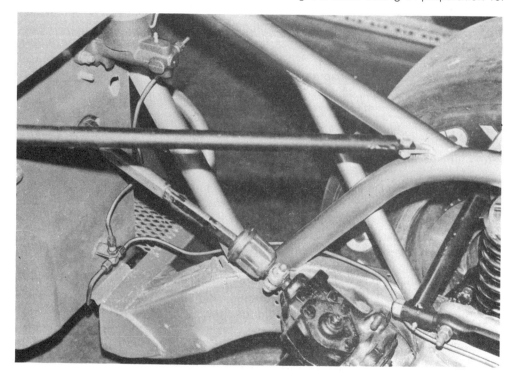

the flare, use a tubing cutter and not a hack saw. The hack saw leaves rough edges and burrs on the tubing, and distorts the diameter and shape of the tubing end as well.

The steel brake lines should be secured to the chassis in several places along its run with rubber lined Adel clamps. These aircraft-designed clamps prevent the transfer of vibration to the brake lines which could otherwise fracture the line material.

FLEX LINES

To prevent a loss of braking system rigidity and for extra safety, use steel braided teflon-lined hose for the flex lines. This is especially critical if disc brakes are being used on the race car. If the steel braided hose is purchased as surplus material, be sure to determine if the hose liner material is teflon or neoprene. The neoprene material is not designed for use in high pressure applications like the braking system.

To determine the amount of flex line to use at each wheel, jack the car up and let the wheel rest a full droop. Then install a line of adequate length which is not tight. For front wheels, steer the wheel hard to one direction as well as setting at full droop to determine the maximum length of flex line required.

All brake flex lines should be from steel braided teflon hose. Keep flex line secured with Adel clamps.

This plumbing is for the rear end and transmission coolers in a Grand National car to be raced on a road course. The pump motors at the right are the Jabsco Water Puppies.

Even steel braided hose should be protected by rubber grommets where it passes through the firewall.

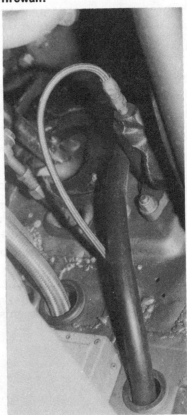

BLEEDING THE BRAKING SYSTEM

To properly bleed any hydraulic braking system, the bleeder valves for the wheel cylinders **must** be at their highest point. On drum braking systems, simple system design and proper installation insures this will be done. But on disc brake systems, it is possible to mount the calipers where the bleeder valves are not at their highest point. If this happens, air will collect in the fluid reservoirs above the level of the bleeder valve opening and remain trapped there during the bleeding process while pressurized fluid is passed through the bleeder valve.

If it should be necessary to mount a caliper in a position where the bleeder valve is not at the highest position, the caliper will have to be removed from its mounting bracket and held in the proper position while it is being bled. If this is done, be sure to place a wooden spacer between the brake pads equal to the approximate thickness of the rotor to prevent excessive piston travel during bleeding.

When first bleeding the braking system after the system's installation and assembly, open the connection between the solid steel lines and the flex lines to bleed the lines first. This prevents trapped air in the lines from being pushed into the wheel cylinders where it may be more difficult to remove. Then reassemble the flex line fittings and bleed the wheel cylinders. When bleeding the front wheels of the car, raise the rear end of the car. Conversely, raise the front of the car when bleeding the rear wheels.

The proper way to bleed brakes is into a bottle with the bleeder hose covered by fluid. This way the system will never suck in any air while bleeding.

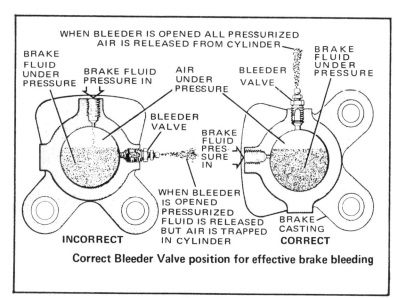

Correct Bleeder Valve position for effective brake bleeding

Drawing courtesy of Halibrand Engineering, a leading maker of race car chassis components.

A good idea for a disc brake master cylinder is to fashion an extra fluid reservoir from another master cylinder. Disc brake systems require more fluid movement than drum systems. The hydraulic clutch slave cylinder and reservoir at the right is from a Datson 280-Z.

STEEL BRAIDED HOSE SPECIFICATIONS

Dash Size	Liner	Hose i.d.	Hose o.d.	Maximum operating p.s.i.	Burst p.s.i.	Min. bend radius
— 3	neoprene	.156	.38	1,000	2,000	1.8''
— 3	teflon	.125	.25	1,500	12,000	1.5''
— 4	neoprene	.219	.44	1,000	2,000	2.0''
— 4	teflon	.190	.33	1,500	12,000	2.0''
— 5	neoprene	.281	.48	1,000	1,700	2.3''
— 6	neoprene	.344	.55	1,000	1,700	2.5''
— 8	neoprene	.438	.64	1,000	1,250	3.5''
—10	neoprene	.562	.80	1,000	1,250	4.0''
—12	neoprene	.688	.94	1,000	1,000	4.5''
—16	neoprene	.875	1.16	750	1,000	5.5''
—20	neoprene	1.14	1.44	500	750	8.0''
—24	neoprene	1.38	1.70	500	750	9.0''
—32	neoprene	1.78	2.02	200	700	12.5''

SUGGESTED USEAGE SIZE OF STEEL BRAIDED HOSE

Use	Liner	Size
Brake lines	teflon	— 3
Fuel lines	neoprene	— 8
Split log to carb fuel bowls	neoprene	— 6
Fuel pressure line	neoprene	— 3
Oil pressure gauge line	neoprene	— 3
Oil scavenge pump lines	neoprene	—10
Dry sump pressure line	neoprene	—12
Dry sump reservoir tank breather	neoprene	—12
Oil cooler lines	neoprene	— 8
Radiator hose	neoprene	—12

A complete discussion of braking system component choice, fluids, hardware, hydraulic system basics, high performance lining, master and wheel cylinders, proportioning, as well as problem solving and system improvement is contained in the book "Race Car Braking Systems" by Steve Smith Autosports. It can be ordered from the publisher under stock #S107.

These are the patented Econo-Seal hose ends by Earl's Supply. A very handy product.

Lubrication

OIL BASICS

Oil has a twofold purpose of lubricating and cooling. In order to choose an oil which will do both of these jobs compatibly and adequately, there are many varying properties you should know about. These properties will affect the manner in which a particular oil protects moving parts, carries away heat, and the residue which it leaves. These important properties include:

Viscosity — which is a measurement of how a particular oil maintains its physical properties under extremes of heat and pressure. High viscosity oils are relatively thick, and offer a resistance to flowing. Low viscosity oils are relatively thin and offer easy flow. The viscosity of an oil effects the oil film thickness which protects rotating and moving parts. Too thin a viscosity will not provide an adequate lubricating film which prevents friction, heat, wear and galling. On cylinder walls in an engine, too thin a viscosity will result in inadequate ring sealing and ultimately high oil consumption. Too heavy an oil viscosity can present as many problems as too thin. The too-thick oil will fail to provide an adequate protective oil film in small clearances, and it will not provide enough heat absorption to conduct away excessive heat.

Viscosity index — is a property which differs from the straight viscosity. While viscosity is the measurement of the pouring ability of an oil at one specific temperature, the viscosity index recognizes the fact that oils change in viscosity properties as they heat up or cool down from that one specific temperature. Some oils will radically thin at much higher temperatures and radically thicken at much cooler temperatures, while other oils will remain more stable and keep basically the same properties through an extreme of temperature ranges. Quite naturally it is easy to see that an oil which dramatically changes viscosity through a range of temperatures is going to cause greater wear of moving parts and great oil consumption.

Viscosity improvers — are additives which make multi-viscosity oils possible. They are made from "polymers" which have molecules that expand and contract in the opposite manner as oil molecules. In other words, under high heat polymer molecules are expanding or getting larger, and under low temperatures they contract or get smaller. What this means in terms of viscosity is that a base of SAE 20-grade oil can be used, with a polymer added that makes it an SAE 10-grade under cold starting conditions, and an SAE 40-grade oil under racing conditions. This is the basis behind a 10W-40 oil. In theory this sounds good, but in actual practice in a racing engine, the polymer additives are not behaving in the same manner as the oil. Under high temperatures and loads the polymers break down after extended use, thus reducing the effective viscosity of the oil. It leaves your engine operating with an SAE 20-grade oil which is undesirable. You will see signs of it in decreased pressure, increased temperature, greater consumption and accelerated wear (especially to rings, bearings and valve stems). There are some good multi-viscosity oils on the market, such as Union's 20W-50 Racing Oil, but be sure to change it after each weekend of racing or after a long race.

Left, no matter if it is the oil filter, oil cooler, or oil lines, be sure all components in the oil system are filled and primed before the engine is first started so you don't pump air to the main bearings. Below, special baffling is required in a wet sump oil pan when a race car switches from oval tracks to a road course.

Oxidation — is an enemy of oil. It happens to the oil through its normal contact with oxygen **at any temperature**. Oxygen molecules are blended with the oil molecules which form corrosive acids. These acids cause wear to metal parts by an etching process. In addition, oxidation causes oil to thicken. While oxidation can take place at any temperature, it is greatly accelerated by high temperatures. This is why oil is recommended to be maintained at 200 degrees F for best results. An oil which has been permitted to get up over 350 degrees F is virtually worthless in a racing engine.

Oil base stock — Any engine oil which you buy in a can is a finished product containing a base petroleum stock with additives thrown in to help the oil achieve certain goals. An engine oil is only as good as the base stock from which it is refined. There are two types of crude oil, based on their molecular structure: paraffinic and naphthenic. The difference between these two types is critical in terms of engine oil performance. The paraffinic molecule has a greater **natural** resistance to oxidation and water. Paraffin oil also provides higher film strength and greater cohesion. The molecular structure of naphthenic crudes, by comparison, is not as resistant to oxidation and water. They oxidize much more readily, building their acid content and becoming corrosive much sooner than paraffinic oils. Engine oils which are commonly referred to as "Pennsylvania Grade" oils are of the paraffinic base, and are more desirable for use in a racing application.

Ash — is a residue which creates deposits which cause preignition. The ash formation occurs from the introduction of anti-corrosive additives and long-life additives in engine oils. High performance racing oils can be purchased in an **ashless** formulation. Buy it.

A recommendation for oil — Taking into consideration all of the properties we have discussed, the perfect oil for a racing engine should be an SAE 40 grade, ashless, paraffinic base, kept to an ideal operating temperature of 200 degrees F. If the operating temperature does not reach 200 degrees F consistently, switch to SAE grade 30 oil. If the oil temperature exceeds 230 degrees F consistently, or if the oil pressure is not up to 55 to 65 pounds, DO NOT switch to a heavier oil than SAE 40 grade. That will only compound your problems. Instead go looking for the **source** of your problems.

Oil additives — Don't be swayed by anybody's sales pitch or decals. Run oil, and only oil for a race. When the engine is new, it should be run-in with an addition of General Motor's EOS (Engine Oil Supplement). But don't go racing with it. It contains calcium and barium, which will leave preignition-causing deposits.

ENGINE OIL COOLING

The ideal engine oil operating temperature, as we said before, is right around 200 degrees F. If your oil temperatures consistently run up in the 230 to 250 degree F range, an engine oil cooler is in order. It will not only help the oil cooling efficiency, but also the water cooling efficiency as the two are highly inter-related (see the Cooling Chapter elsewhere in this book for more details).

The oil radiator, or more properly termed heat exchanger, should be chosen on the basis of its capacity, restriction and fins-per-inch count. The capacity is important to insure the

The Harrison/GM oil radiator, part number 3157804, available from your local Chevy dealer. See text for a similar part available from Ford dealers.

Oil cooling radiators placed inside the race car do not get the job done efficiently because inside the car is a dead air spot.

oil spends a reasonable amount of time in the cooler so heat can be adequately exchanged. A small, low capacity cooler will pass the oil through quickly without allowing an ample opportunity for oil cooling. The restriction in a cooler usually is the result of the oil being flowed through very small diameter tubes. This is bad — you already have enough built-in restrictions in the oil system without adding any more. The fins-per-inch count has a bearing on the heat exchange capacity of the unit. The more fins, the better the heat exchange. A good oil cooler will have 13 to 15 fins per inch.

Many heavily advertised commercial oil coolers do not have the capacity and fins-per-inch to get the job done. They do cool the oil — usually resulting in a temperature drop of only a few degrees from entrance to exit. This is not worth the expenditure for the unit.

A highly recommended oil cooler is the Harrison part number 3157804, available from your local Chevrolet dealer. Another very similar cooler is Ford's part number C90Z-6A642-A. They are big, and expensive. But, they get the job done, dropping the oil temperature as much as 30 degrees from entrance to exit (exact amount depends on air flow routing and ambient temperature), with a minimum of restriction. When used with a dry sump system, they are guaranteed to withstand the high pressures of the system.

When using an oil cooler, plan your plumbing route carefully to avoid long runs of hose, an abundance of bends or elbow fittings. All of these serve to restrict oil flow. If yhou are using two oil coolers, plumb them in parallel, never in series. This prevents a loss of oil pressure.

When installing the oil cooler and its routing lines, be sure to fill them **completely** with fresh oil. Failure to do so will

mean you will be feeding the main bearings with air in the first few minutes of engine operation.

Mount the engine oil cooler in an area where it is assured of a constant supply of fresh, cool outside air. Even better, build a ducting from aluminum sheet material (such as we describe for the water radiator in the cooling chapter) to route air directly into the cooler to assure maximum efficiency from it. Completely protect the cooler from flying

Oil should enter the cooler from the lowest point in order to force air out of it and prevent air lock.

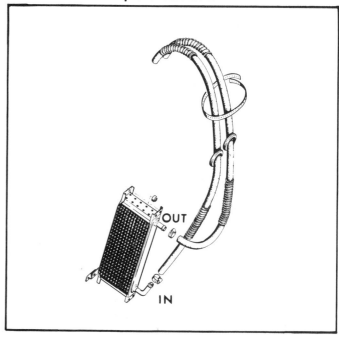

The oil cooler should be well protected with a mesh screen so stones and debris do not nick it. Housed in the aluminum bracket to the right of the oil cooler is the remote oil filter.

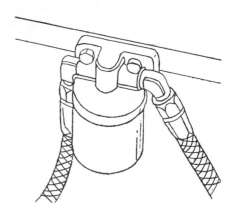

When you attach a remote oil filter to the roll cage in the engine compartment, weld a bracket to the cage tube, then bolt the filter adaptor to the bracket. Never bolt the filter adaptor right to the tube — drilling holes in it will severely weaken the tube.

rocks and debris — front and rear — with a mesh screen guard. Don't ever mount an oil cooler inside a race car — even if the car has no windshields — because the inside of the car is a dead air spot. It will not get the air flow which is required for maximum efficiency. Likewise, don't mount the oil cooler in front of or behind the water radiator. Neither one of the radiators can afford to be fed with heated air. The oil cooler should be plumbed so that the oil enters from the lowest point in order to force air out of the cooler so there will be no air lock.

Whether you are using a dry sump or wet sump oiling system, you should use an in-line aircraft-type screen filter just before the oil cooler. This helps to de-aerate the oil, which will aid in cooling efficiency, plus it prevents trash collecting in the cooler should you blow an engine. One filter is sold by Aviaid — the oil pan people. Another type to be considered for use is the Mecca Engine Condition Inspection oil filter system sold by Autoworld in Scranton, Pennsylvania. It is a cast metal body which incorporates a removable filter and de-aerating screen. It has an excellent filtering element which will more than suffice to replace any paper-element oil filters, with a minimum of restriction.

OIL FILTERS

Stock passenger car cannister oil filters are designed to bypass oil flow at a preset pressure. So, under high pressure you have a chance of circulating unfiltered oil. There are special oil filters made by STP, Purolator and Fram which strive to improve this situation. All are sold under the part number of HP-1. They will screw in to any remote filter adaptor. Be sure to use the high performance filter.

The Mecca E.C.I. filter we described previously is a better alternative to the conventional paper element oil filter.

OIL GAUGES

The engine oil pressure gauge should be read directly from the engine — not the remote filter or radiator. This is the only way to directly monitor situations which directly affect the healthy operation of your engine. A minimum of 3/16-inch line should be used to feed the pressure gauge to make it more sensitive.

The oil temperature should be monitored where the oil re-enters the engine from the cooler — after all, this is the temperature of the oil which the main bearings see.

This is the Mecca Engine Condition Inspection oil filter sold by Auto World.

PLUMBING THE OILING SYSTEM

At the minimum, you will probably be plumbing a remote filter and an oil cooler. To do the job right, use a steel braided hose (neoprene liner is fine) with a minimum inside diameter of approximately ½-inch (which would be the dash eight size).

The proper routing of oil lines for a wet sump engine is from the engine to the filter, to the cooler, and back to the engine. It is preferable for the oil to be filtered before cooling so as much debris as possible can be kept out of the cooler.

Before first installing new hoses and a new cooler, be sure to pour solvent, then clean oil, through them to clean them. Don't assume they are clean right out of the box. It could be a costly mistake.

THE DRY SUMP SYSTEM

A dry sump system is costly. No getting around that. For any type of racing, short track or long, it is used mainly as a dependability factor. It is also a great aid in allowing an engine to be placed lower in the chassis, which in turn lowers the vehicle's center of gravity heighth for an improvement in handling. The dry sump gives less horsepower loss through hydraulic frictional drag by preventing some oil wind-up around the crankshaft. It allows cooler engine operating temperatures by circulating a very large volume of oil. And finally, it takes some strain away from the camshaft by not having to turn the driveshaft on the stock-type oil pump.

For the most part, most of the advantages associated with the use of a dry sump system can be overcome by careful and proper design of components used in a wet sump, such as a good windage tray and crankshaft scraper, good pan design, and the use of an Accusump pressurizing device (write to Autoworld in Scranton, Pennsylvania, for more details).

For more information on proper dry sump system design and installation, see our other books, "Racing Engine Preparation" and "Racing The Small block Chevy."

OTHER LUBRICANTS

When you need grease to lube chassis parts, bearings, etc., don't go down to the local service station and have them give you a coffee-can full of "whatever is handy." Today's world of greases and lubricants is very sophisticated with many greases formulated for special applications. One of the best breakthroughs in lubrication technology for racers is the disc brake being made standard on passenger cars, because special high temperature, heavy duty service wheel bearing greases had to be formulated. This matches the service racers require for heavily loaded wheel bearings.

For grease in wheel bearings, universal joints, steering linkage and suspension linkages, use Valvoline's Val-Lith EP No. 2, or Union's Unoba EP-2.

For the transmission and rear end, we have seen good results with three different gear lubes: Phillips 66 # GL59081 SAE Grade 120 gear lube, Valvoline's #829 SAE 85W-140 gear lube, and Lubrication Engineers' Almasol #601 SAE Grade 90 gear lube.

We have been especially impressed with the results of Lubrication Engineers' Almasol gear lube. We have seen quick change rear ends using the Almasol which have had hard use on race tracks for over two years, and the gears and bearings exhibited almost no signs of wear! The Almasol is an industrial gear lubricant which was especially designed for uses where gears are continually subjected to overloading and overspeeding. Without special lubricants, these gear sets would experience accelerated wear or breakage. Almasol is actually the name of Lubrication Engineers' exclusive wear reducing agent which they add, which coats all metal surfaces with a microscopic layer that separates metal surfaces should the oil film rupture under heavy loading (like torque shock loads under acceleration). The unique Almasol material possesses tremendous load carrying capability, is unaffected by temperatures up to 1900 degrees F, and is impervious to corrosive acids attack. The base stock for this gear lube is a refined paraffin oil, which has the highest **natural** viscosity index.

The Roll Cage

THE ROLL CAGE AND STEEL STRUCTURAL TUBING

The roll cage structure occupies the majority of round steel tubing in a stock car, whereas the frame structure uses the greatest portion of square and rectangular steel structural tubing. But before we get into the actual fabrication of these component assemblies, a background in steel material composition and properties is in order.

STEEL GRADE NUMBERING

A four-numeral series is used to designate graduations of chemical compositions of steels and alloy steels, such as 1010, 1018, 1020, 1024, 4130 and 4340. The first two digits of the numerals represent the various grades of materials, such as:

1000 — nonresulphurized carbon steel grades
1100 — resulphurized carbon steel grades
1200 — resulphurized and rephosphorized carbon steel grades
4100 — alloy steel with .50% to .95% chromium and .12% to .30% molybdenum
4300 — alloy steel with 1.83% nickel, .50% to .80% chromium and .25% molybdenum

The last two digits of the four numeral series indicate the approximate middle range of the carbon content. For example, in the designation 1035, 35 represents a carbon range of .32 to .38 percent carbon (35 is in the middle of 32 to 38).

GENERAL PROPERTIES OF METALS

There are seven basic properties applied to all metals and materials which greatly influence the choice of a particular metal for a particular job.

Brittleness — A brittle metal will crack or break when subjected to bending or deforming. A brittle metal cannot be used where shock or torsion loads are expected. The higher the carbon content of a steel, the more brittleness it has to it.

Ductility — A ductile metal can be easily bent, drawn or twisted into shape without breaking. Malleability is similar to ductility. Low carbon content steels (sometimes called "soft" steels) have great ductility.

Elasticity — This is the ability of a metal to return to its original shape after deflection or bending of it. It will not stay permanently deformed after a load distorts it. Elasticity is an important property in axles and springs.

Density — Density is the weight of a given metal per cubic inch. Alloy steel is much more dense than aluminum.

Toughness — This means a metal will withstand tearing or stretching without breaking.

Conductivity — Conductivity is the ability of a metal to carry or transmit heat and electricity.

Hardness — Hardness is the ability of metal to resist cutting, permanent deformity, wear and abrasion. Hardness is increased in a metal by coldworking. Hardness is increased in steels by the addition of carbon (among other elements).

HARDNESS TESTING

To determine the hardness of a metal, a testing device called a Rockwell hardness tester is used. It is almost similar in appearance to a valve spring rater. It has a diamond tip penetrator which is forced into the surface of the metal being tested. The depth of the impression of the tip, compared to the load being applied on the tip, gives a scale reading. There are two Rockwell scales, a B scale and a C scale. The B scale is for softer metals, the C scale for the harder ones. Most generally, the C scale is the most widely used one for any hardness tested parts being used in a racing vehicle. The scale ranges from C-20 (soft) to C-70 (very hard).

BASIC STRESSES APPLIED TO METALS

There are five basic stresses which are applied to metals and materials, and the metal content as well as material shape is important in the choice of materials to resist these stresses.

Bending — Bending is the application of a deflecting force on a structure. It is also known as buckling. With round steel tubing, the greater the length of tubing the less resistance it has to buckling or bending. An increase in round tubing diameter can help the length of tubing resist bending loads. Bending stiffness increases with the cube of the diameter increase.

Compression — This is the application of two forces towards each other, applied on opposite ends of the structure. Compression is the opposite of tension or tensile strength. Round tubing resists compression loads better than square or rectangular tubing.

Shear — is the application of a straight-on cutting force. It can be pictured as a piece of metal being cut by a scissors.

Tension — Also called tensile strength, this is the resistance of a metal to stretching forces which want to pull it apart.

Torsion — This is the application of a twisting force to the metal section. An increase in round tubing diameter increases resistance to torsional deflection.

ROLL CAGE MATERIAL CHOICES

There are four basic methods of producing round steel tubing — seamless, D.O.M., cold drawn butt welded, and electric resistance welded. There is a large difference in the strength, quality and price of the tubing produced by these four different methods. And to achieve a top quality cage, price cannot be the only determining factor in choosing material.

Seamless tubing is produced by a method which is basically a forging operation. The metal is worked from the inside as well as the outside. This results in a refined grain structure as well as a uniform grain flow. This type of tubing contains a slight spiral twist which gives added strength and ductility. Hot finished seamless tubes have a surface comparable to hot rolled bars. Cold finished tubes have the advantage of a better finish, closer dimensional tolerances and higher strength.

Drawn Over Mandrel (D.O.M.) steel tubing (also known as welded and drawn) is formed from strip steel and is electric resistance welded. The welded tube is then normalized and cold drawn to a smaller diameter and thinner wall. The cold drawing process works the weld area to produce a sound, dense, homogeneous structure. The weld line disappears, causing the tube to be virtually seamless. Each length of tubing is subjected to non-destructive testing to insure product quality. Because of the cold working process, D.O.M. tubing is superior in strength characteristics to seamless tubing.

Cold drawn butt welded tubing is cold drawn to size from hot rolled, continuous welded material. The cold drawing operation is identical in every respect with the method used in producing seamless tubing. But, cold drawn butt welded tubing is drawn on the outside only, leaving a large tolerance variance in the wall thickness.

Electric resistance welded tubing is manufactured by forming flat rolled steel into a tubular shape and welding the edges. The flash is always removed from the outside of the tube. The inside welding flash is usually controlled to a heighth of .010-inch.

ROLL CAGE FABRICATION

The roll cage is designed and built for two purposes — to afford the driver the best possible protection, and to add to the chassis stiffness and rigidity. But don't lose sight of the primary function of the cage: safely protecting the driver. Material diameter and wall thickness, and welding quality should all be considered when building the roll cage. Above all else, the minimum requirements of chassis construction of any racing association should be strictly adhered to.

This is a good example of a very well-triangulated roll cage and chassis, built with a selective use of roll cage tube sizes to keep weight down.

EFFECTS OF COMMON ALLOYING ELEMENTS IN STEEL

By definition, steel is a combination of iron and carbon. Steel is alloyed with various elements to improve physical properties and to produce special properties, such as resistance to corrosion or heat. Specific effects of the addition of such elements are outlined below:

Carbon (C), although not usually considered as an alloying element, is the most important constituent of steel. It raises tensile strength, hardness and resistance to wear and abrasion. It lowers ductility, toughness and machinability.

Manganese (Mn) is a deoxidizer and degasifier and reacts with sulphur to improve forgeability. It increases tensile strength, hardness, hardenability and resistance to wear. It decreases tendency toward scaling and distortion. It increases the rate of carbon-penetration in carburizing.

Phosphorus (P) increases strength and hardness and improves machinability. However, it adds marked brittleness or cold-shortness to steel.

Sulphur (S) improves machinability in free-cutting steels, but without sufficient manganese it produces brittleness at red heat. It decreases weldability, impact toughness and ductility.

Silicon (Si) is a deoxidizer and degasifier. It increases tensile and yield strength, hardness, forgeability and magnetic permeability.

Chromium (Cr) increases tensile strength, hardness, hardenability, toughness, resistance to wear and abrasion, resistance to corrosion and scaling at elevated temperatures.

Nickel (Ni) increases strength and hardness without sacrificing ductility and toughness. It also increases resistance to corrosion and scaling at elevated temperatures when introduced in suitable quantities in high-chromium (stainless) steels.

Molybdenum (Mo) increases strength, hardness, hardenability and toughness, as well as creep resistance and strength at elevated temperatures. It improves machinability and resistance to corrosion and it intensifies the effects of other alloying elements. In hot-work steels, it increases red-hardness properties.

Tungsten (W) increases strength, hardness and toughness. Tungsten steels have superior hot-working and greater cutting efficiency at elevated temperatures.

Vanadium (V) increases strength, hardness and resistance to shock impact. It retards grain growth, permitting higher quenching temperatures. It also enhances the red-hardness properties of high-speed metal cutting tools and intensifies the individual effects of other major elements.

Cobalt (Co) increases strength and hardness and permits higher quenching temperatures. It also intensifies the individual effects of other major elements in more complex steels.

Aluminum (Al) is a deoxidizer and degasifier. It retards grain growth and is used to control austenitic grain size. In nitriding steels it aids in producing a uniformly hard and strong nitrided case when used in amounts 1.00% - 1.25%.

Titanium, Columbium, and Tantalum (Ti, Cb, Ta) are used as stabilizing elements in stainless steels. Each has a high affinity for carbon and forms carbides, which are uniformly dispersed throughout the steel. Thus, localized depletion of carbon at grain boundaries is prevented.

Lead (Pb), while not strictly an alloying element, is added to improve machining characteristics. It is almost completely insoluble in steel, and minute lead particles, well dispersed, reduce friction where the cutting edge contacts the work. Addition of lead also improves chip-breaking formations.

What elements are added to steels to change their properties? The table above — courtesy of Earle M. Jorgensen Co. — explains.

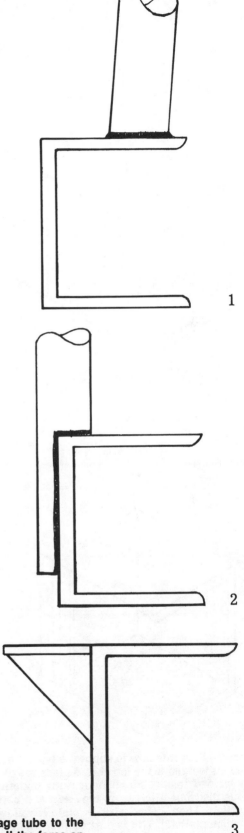

1, 2, and 3 illustrate three methods of attaching a roll cage tube to the frame. In 1, welding the tube on top of the channel puts all the force on the top unsupported flange — very poor. In 2, the tube is sectioned to fit over the frame channel — good. Ties in full strength of frame rail. In 3, a pad is welded to the outside of the channel. If the pad is well supported by gussets this is a good method.

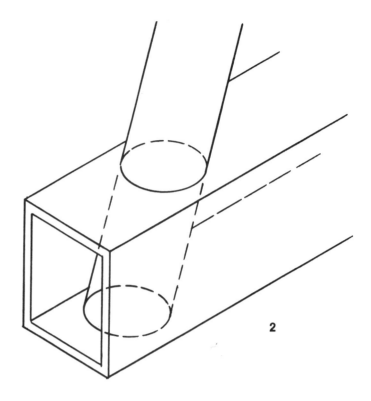

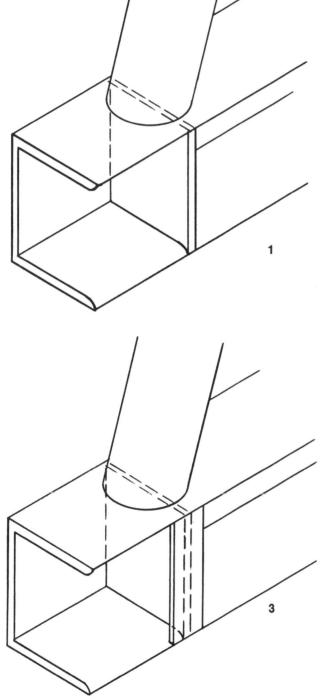

1 and 3 illustrate how to support a typical frame rail when a tube is welded to the top of it. A piece of 3/8-inch steel plate is welded inside the channel right underneath the tube. This distributes forces on top, side and bottom of channel section. Then a short strip of plate is welded over front of section plate [3]. This is the best method of attaching a tube to the frame rail. In 2, this method of attaching the tube to the top and bottom of a unibody frame rail should be used to prevent overstressing the thin gauge material of the rail.

Quality welding is a key to a quality cage. It definitely must be arc welded, with the MIG or TIG arc processes preferred (see welding chapter for a definition). But, a good job with a stick arc is also acceptable.

The other key to a quality cage is the material selection. 1010, 1018, 1020 and 1025 mild steel seamless or D.O.M. tubing can be used, or 4130 chrome moly seamless tubing sometimes is used. If the choice is one of the mild steels (as it should be), use nothing more brittle than 1018. There is a wide difference in ductility and brittleness between 1010 and 1020 material. 1010 is soft, energy absorbing and very workable. 1020 can be quite brittle. The 1018 material seems to be a good compromise for roll cages and race car construction, and is the material preferred by most professional race car constructors. In choosing between seamless and D.O.M. tubing, the D.O.M. is considered superior. The use of 4130 chrome moly steel is discouraged because the chance of weld embrittlement is great with it. If 4130 is used for the entire cage (and by all means don't mix 4130 with mild steels), the **entire** roll cage structure should be stress relieved (or normalized). If this is not done, the structure won't even be as strong as a comparable mild steel cage, because of the weld embrittlement problem.

All cage tubes should be bent in a mandrel tubing bender, not a compression or crush-type bender. A mandrel bender produces a smooth, wrinkle-free bend. A wrinkled bend will deflect or completely fail under high stresses. A compression bender makes the tube diameter at the bends smaller than the rest of the tubing diameter, which means the highly stressed bends will be the weakest points in the cage with a

Square tubing should never be used as an integral part of the roll cage, such as this stressed member here. It does not have the strength of round tubing.

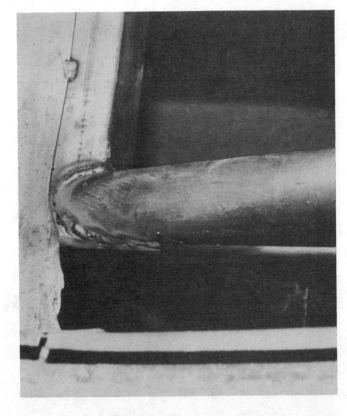

Roll cage tubes which attach to the frame at an angle can be attached in two ways. Above, a welded gusset box or torque box is used. At right, the tube is cut at an angle to mount directly to the frame. The method at the right is more time consuming [because of computing the angle], but it is stronger because it spreads the loads in the tube and the chassis in a wider area.

compression bender.

The joints of two or more intersecting tubes must be shaped and fitted together. Do not use weld material as a gap filler. This definitely will affect the strength of the joint.

In fitting two intersecting tubes together, the open-ended tube should be notched so that it extends into the main tube to a depth of at least one half the diameter of the main tube. This is to give a sufficient strength to the welded joint.

When the major tubes of the cage contact the frame rails, a simple butt weld from the tube to the frame is not sufficient (except in the case of cage tubes being welded to rectangular tubing frame rails). The tube should be

extended through the top frame surface and to the bottom surface. This is especially important on any unibody structures. The tube should be completely welded to both surfaces. This is much stronger and stiffer than a simple butt weld. Additionally, on unibody cars, the tubes must be welded to major structural members such as rocker panels, and not to thin flexible sections such as the floor pan.

The roll cage tubes should be welded to the car's body sheetmetal (such as window posts, door posts, etc) to increase the structural stiffness of the vehicle. This integrates the chassis and body into one unit.

When the cage is being built and fit, cut and tack-weld

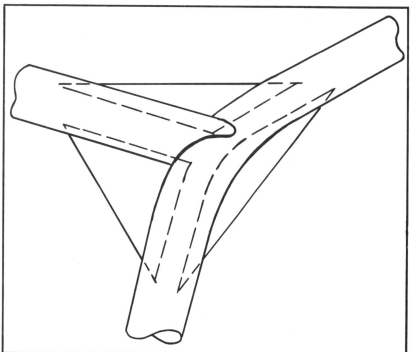

Left, these gussets are used to strengthen highly stressed areas, and to keep intersecting bars from separating under impact or high stress.

Below, the straight door bars may not look as neat as the curved bars, but they sure add a lot more strength to the complete cage.

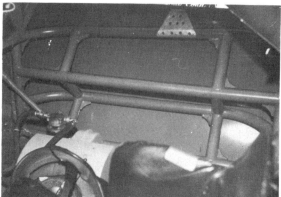

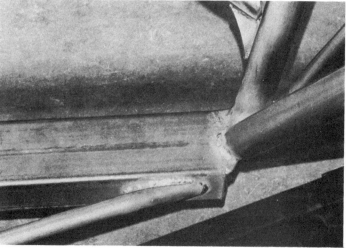

Above left, the upper main cage hoop is curved to fit the contour of the sheetmetal. When welded to the sheetmetal it adds strength to the total chassis. Above right, a beautifully triangulated intersection. Left, one method of attaching the body to the frame and tying in a cage tube. Adds strength.

These three photos illustrate a well-triangulated chassis, designed by Paul Lamar. All bars are straight except for the door bars, and one radius in one tube. Short sections of straight tubes add incredible strength to the chassis.

one piece at a time. This insures a better fit and less chance for errors. The entire cage should be tack-welded together before the finish welding of all seams, and the finish welding should be done alternately between the front, central and rear portions of the cage. All of these steps are taken to eliminate heat warpage of the chassis structure from welding.

To shape the cage to the body, set the gutted-out body on the chassis in order to trial fit the major cage structure. Tack the main cage members into place, then remove the body to finish the welding and supporting bar fitting.

ROLL CAGE GUSSETS

Roll cage gussets should be used where the roll cage structure attaches to the frame, and where major components of the cage join together. The purpose of the gusset is to spread loads and stresses out through the cage tubes further than just at the welded joints. The most common gusset used in roll cages is made from a 2-inch by 2-inch by 2-inch triangular piece of mild steel sheet, 1/8-inch thick. A more effective type of gusset is the V-gusset. It is made by folding a piece of .049-inch thick sheet steel in half

around a 3/4-inch diameter rod, then cutting the folded piece in a triangular shape on the open ends. Weld the folded gusset in place like a traditional flat gusset. The result is a much superior load carrying gusset.

WHICH MATERIAL SIZES

Which tubing diameters and wall thickness should be used for each particular application in a roll cage structure is always a puzzling question for many builders. The important aspect with material choices is to achieve maximum strength with a minimum of weight. To achieve those goals, we offer the following recommendations:

Assume we are building a roll cage of the design we have illustrated. If the entire structure were built with 1-3/4-inch O.D., .095-inch wall tubing, the 153 feet of tubing used would weigh 257 pounds. However, substitutions can be made to both save weight and increase strength.

The four-point central cage bay (coded A on the drawing) takes 33.5 feet, the largest single structure in the cage. Most racing associations require 1-3/4-inch O.D. tubing with .095-inch wall here, so we agree and use that tubing designation. 33.5 feet of this tubing weighs 56.2 pounds.

The "X" section (B) in the main hoop behind the driver uses 4.6 feet of tubing. Use 1-3/4-inch O.D. with .065-inch

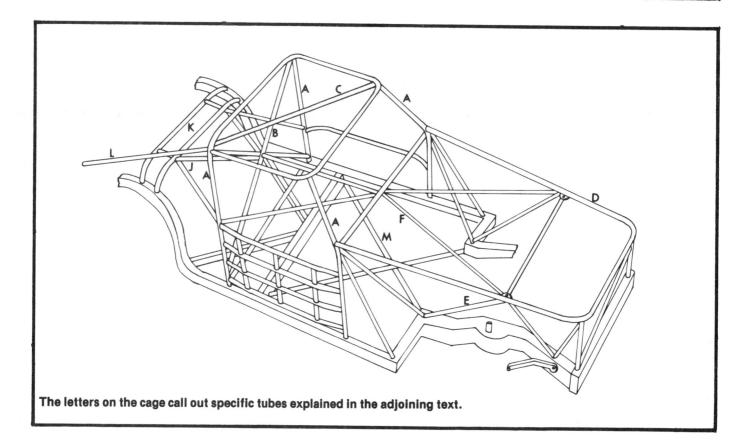

The letters on the cage call out specific tubes explained in the adjoining text.

wall. Total weight is 5.4 pounds.

The diagonal in the top of the cage (C) uses 4.8 feet and should be 1-3/4-inch O.D. with .065-inch wall tubing. Total weight is 5.6 pounds.

The front bay main hoop (D) requires 13 feet of material, and it should be 1-3/4-inch O.D. with .065-inch wall tubing, giving a total weight of 15.2 pounds. The front bay hoop bracing (E) should be the same material, it takes 8.3 feet of tubing and weighs 9.6 pounds.

The front bay top diagonal (F) should use four feet of 1-1/2-inch O.D., .065-inch wall material, for a weight of four pounds.

The door bars use 13.3 feet of tubing on each side of the car. On the left side, 1-3/4-inch O.D., .095-inch wall material should be used, for a weight of 22.1 pounds. On the right side, 1-3/4-inch O.D., .049 wall material can be used, for a weight of 11.8 pounds (notice the amount of weight saving for a .045-inch reduction in wall thickness for the same length of material).

The seat frame (I) uses seven feet of material. It can be 1-1/2-inch O.D., .065 wall, for a weight of seven pounds.

The bracing from the main hoop to the rear spring buckets (J) uses 12 feet of material. It should be 1-3/4-inch O.D., .065-inch wall, for a weight of 14 pounds.

For the bracing between the rear spring buckets (K), use three feet of 2-inch O.D., .049-inch wall tubing, which will weigh three pounds. This size material was chosen to resist compressive and torsional loads.

The parallel braces from the top of the main cage hoop to the rear bumper (L) is a very long span of tubing, and so need an increase in diameter to resist bucking loads in the event of a rear end collision. 2-inch O.D., .065-inch wall tubing should be used here. There are 13 feet of material used, and the total weight is 17.5 pounds.

The "X" member between the frame rails (M) should utilize 1-3/4-inch O.D., .065-inch wall tubing. There are 13 feet in this component, for a total weight of 15.2 pounds. A 2-inch O.D., .049-inch material would be better in this location to better resist torsional and bending loads, as well as reduce weight. But the 1-3/4-inch material was chosen as a compromise to provide more component clearance in the under-the-floorpan area.

The total weight of our selectively designed cage is 191 pounds, a total saving of an incredible and very important 66 pounds over the cage built with just 1-3/4-inch O.D., .095-inch wall tubing!

In any position in the above discussion where our recommended material sizes conflict with materials required by the rules of your racing association, we recommend you make the appropriate adjustments in the interest of safety and legality.

Never drill any holes in the roll cage structure to mount brackets or pop-rivet sheet metal to the tubes. The holes cause stress areas and weaken the tubes. It is best to weld tabs to the cage tubes on which brackets can be attached. This is more time consuming, but it is the only way to do the job right.

WEIGHT OF STEEL ROUND TUBING*

O.D.	WALL	I.D.	WEIGHT PER FOOT
.75	.035	.680	.2673
.75	.065	.620	.4755
.75	.125	.500	.8344
1.0	.049	.902	.4977
1.0	.065	.870	.6491
1.0	.095	.810	.9182
1.25	.049	1.152	.6285
1.25	.095	1.060	1.172
1.25	.120	1.010	1.448
1.5	.049	1.402	.7593
1.5	.065	1.370	.9962
1.5	.083	1.334	1.256
1.5	.095	1.310	1.426
1.5	.120	1.260	1.769
1.75	.049	1.652	.8902
1.75	.065	1.620	1.170
1.75	.083	1.584	1.478
1.75	.095	1.560	1.679
1.75	.120	1.510	2.089
2.0	.049	1.902	1.021
2.0	.065	1.870	1.343
2.0	.083	1.834	1.699
2.0	.095	1.810	1.933
2.0	.120	1.760	2.409

*All round steel tubing such as 1010, 1018, 1020, etc., and all types of manufacturing processes.

MECHANICAL PROPERTIES COMPARISON OF ROUND STEEL TUBING*

MATERIAL	TENSILE STRENGTH	YIELD POINT	ELONGATION
1018 hot finished seamless	60,000 p.s.i.	35,000 p.s.i.	30%
1018 cold drawn seamless	80,000p.s.i.	60,000 p.s.i.	15%
1018 D.O.M. welded and drawn	80,000 p.s.i.	70,000 p.s.i.	15%
1010 cold drawn butt welded	65,000 p.s.i.	50,000 p.s.i.	20%
1010 cold rolled resistance welded	45,000 p.s.i.	35,000 p.s.i	25%
1010 hot rolled resistance welded	48,000 p.s.i.	35,000 p.s.i.	25%
1020 cold rolled resistance welded	50,000 p.s.i.	38,000 p.s.i.	22%
1020 hot rolled resistance welded	52,000 p.s.i.	38,000 p.s.i.	20%

* Properties refer to material sized up to 2-3/4 inches O.D. by .124-inch wall.

Mounts and Brackets

FABRICATING ENGINE MOUNTS

There are three basic areas of concern when designing and fabricating engine mounts — getting the engine low enough, getting the engine mounted at the proper angles, and allowing enough clearance for headers and steering linkage as well as providing ease of engine installation and removal.

TYPES OF ENGINE MOUNTS

Race cars today are in a period of transition from using passenger car-type of side engine mounts to using solid mounting engine plates which grip the engine at the front and rear of the block. In almost all forms of motor racing, the transition has been made to the solid plate mounts. But in stock car racing, the evolution is just beginning.

Why use the solid plate mounts? For low performance passenger cars, the side engine mounts are quite suitable. But in high performance applications, the side mounts induce cylinder wall flex during torque reaction in the chassis. It is at this time that the engine can ill afford cylinder wall deformation. Gripping the block at its front face and rear face will have minimal effect on block distortion. It also is a far more stable way to mount a heavy-weighted rectangle than to pivot it on its sides. The front and rear plates can be solidly attached to the frame, allowing no engine movement from torque reaction, and thus eliminating problems such as clutch linkage bind. The engine mounting plates also yield much more room on the

sides of the block for header clearance, as well as making the engine installation easier.

If the passenger car-type side engine mounts are used, at least the left side mount should be rubber mounted. The left side mount receives both tension and radial loads under torque reactions. If it were bolted to the block solidly, in time a large piece of material would pull right out of the side of the block. Because the cast iron block is more flexible than the solid mount, it would flex and eventually crack and fail because cast iron is not a malleable material. If the left side mount is rubber, it must have a restraint such as a chain,

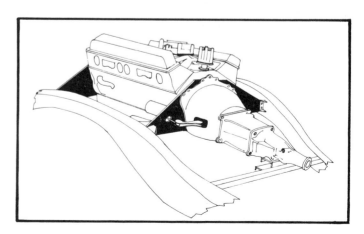

Gripping the engine with solid mounting plates is a much more sensible and efficient method of mounting the engine.

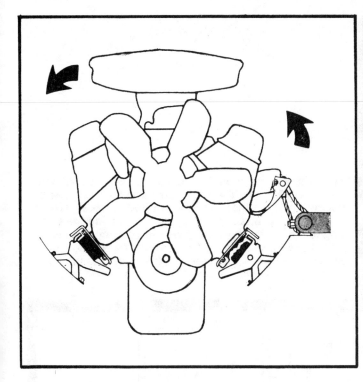

Left, Arrows illustrate torque reaction movement of an engine in the chassis. A tie-down strap is required to prevent broken engine mounts if rubber mounts are used. Above, the solid mounting plates used on the front of the engine in a leading late model sportsman car.

steel strap, heim joint link, or confining bolt to limit its stretch. Otherwise, when subjected to extreme tension it will break and allow the engine to rotate in the chassis using the right mount as a pivot.

FABRICATING THE MOUNTS

For the trial fit of the engine in the chassis, a bare dummy block is needed, with the heads, intake manifold, headers, fuel pump, oil pan, bellhousing and transmission all fitted to it. All of these exterior engine components can and will dictate precise placement of the engine in the chassis.

The most critical clearance on the bottom side of the engine is between the oil pan and drag link. To help easily judge that clearance, braze or tape a piece of 1-inch by 4-inch tubing to the oil pan where the pan would contact drag link. The one-inch direction of the tubing would be placed to create a one inch clearance between the pan and the drag link when the engine is set down on it.

For measurement purposes, block the dummy engine into place with boards and scraps of steel to stabilize it. Once the engine is low enough in the chassis and the front engine mounting location has been established, a small floor jack under the tailshaft of the transmission can lift the engine up and down to establish the desired engine mounting angle.

To establish the engine mounting angle, pop the intake manifold off, lay a steel straight edge across the front and rear of the block and use a mechanic's level placed on top of the straight edge. Use it to determine that the engine is angled four degrees downward at its rear (in other words, the front end of the block should be slightly higher than the rear face of the block, to facilitate water flow and eliminate steam pockets in the block and cylinder heads).

The next measurements made for alignment should be those which determine that the engine is placed with the crank centerline (as seen in the top view) parallel with the centerline of the vehicle. We could say parallel with the frame rails, but on some full frame chassis the rails are curved or do not run parallel with each other. The center of the engine need not be the same as the center of the chassis, even though the pinion gear is centered at the rear end. The engine may be offset to the left to get a little more weight bias to the left. But, the engine must be set in the chassis straight, and not have the front of the engine closer

In order to properly fabricate the engine mounts, a dummy engine must be used for fitting. Note the left side offset on this baby.

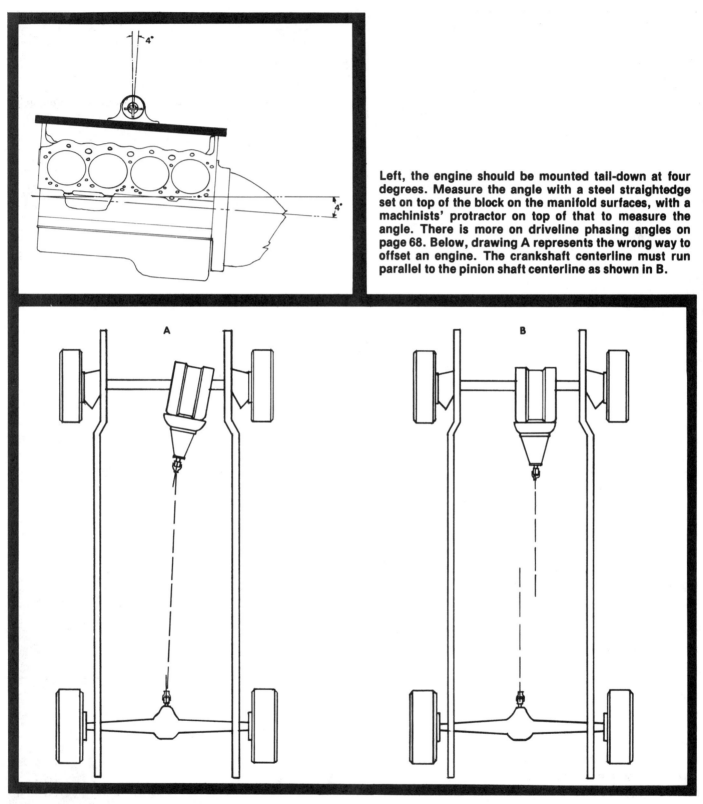

Left, the engine should be mounted tail-down at four degrees. Measure the angle with a steel straightedge set on top of the block on the manifold surfaces, with a machinists' protractor on top of that to measure the angle. There is more on driveline phasing angles on page 68. Below, drawing A represents the wrong way to offset an engine. The crankshaft centerline must run parallel to the pinion shaft centerline as shown in B.

to the left side frame rail than the back of the engine. To check for this, measure from the center of the crankshaft dampener (or the center of the crank hole is if you are just using a block) to the frame rail on each side. If the engine is centered, the measurement will be the same from the crank to each frame rail. If the block is offset to the left 1-3/4-inches, the measurement to the right side rail will be 1-3/4-inches greater than the measurement to the left side frame rail. Make note of these distances. Then measure from the center of the transmission tailhousing to the frame rails on each side. Unless the frame rails are parallel all the way back, the measurements to the frame rails in the back

Left, a well-constructed solid side engine mount. Above a neat triangulated bracket for mounting a six banger. Below, a front mounting plate as used in a Formula car.

will not be the same lengths as in the front. But that is not important. What is important is the difference between the center of the tailhousing and frame rail on each side. It should be exactly the same as the difference between the crankshaft center and the frame rail on each side at front. Assume that we had determined that the front of the engine is offset to the left 1-3/4-inches. Then the measurement from the center of the tailhousing to the left side frame rail should be 1-3/4-inches less than that from the center of the tailhousing to the frame rail at the right. If the engine crankshaft centerline is not placed parallel to the vehicle centerline and pinion gear centerline, severe vibration problems can develop. Once the engine is alligned in the chassis properly, wire or support everything so it does not move, and begin to construct the engine mounts.

Use pieces of light cardboard to make templates for the engine mounts. Do this regardless of whether you plan to use side engine mounts or engine mounting plates. Cut and trim the cardboard, and install the final configuration to the engine and chassis just as if it were the mount. If everything fits, and there is no interference, transfer the cardboard patterns to metal.

In the case of engine mounting plates, the template can be laid directly on top of the sheet material and traced on it. Then the plate can be flame-cut to size and shape. The plates should be fabricated from .1875-inch thick steel sheet, and bolt to a frame rail bracket on each side with at least two 5/16-inch bolts in each bracket. The plates can also be made from .375-inch thick 6061-T6 aluminum, but a longitudinal engine locating device would have to be employed in order to prevent fore and aft engine movement. Any fore and aft movement of the engine would fatigue and crack the aluminum plates.

The rear engine mounting plate is also cut from .1875-inch thick sheet steel, and is designed to sandwich between the rear block face and the bellhousing. It extends to the frame rails on each side, and can either be welded to the frame rails, or bolted to brackets which are welded to the frame rails. In either case of attachment, the plate adds greatly to the chassis torsional stiffness which is very vital in this area. An engine mounting plate placed in this location also takes the load bearing requirements off the transmission tailshaft.

Side engine mounts can be fabricated from square tubing, such as 2-inch by 2-inch, with .095-inch wall. In this

A solid side engine mount made of square tubing. Problem here is access to mounting bolt.

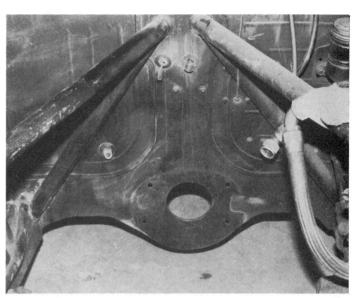

A very competitive late model sportsman car uses this rear solid mounting plate. Firewall is welded to plate, triangulated by cage tubes. Plate is welded to frame rails. Very rigid, and very lightweight. Below, a triangulated tubular engine mount, with rubber cushioning. This type of rubber mount is not prone to rupturing from torque reaction.

case the cardboard template can be used to determine spacing and chassis and block cutting angles for the tubing. If a passenger car rubber engine mount is to be used, bolt it to the block, then fashion the template up to the rubber mount. See accompanying photos for more information on engine mount construction.

THE TRANSMISSION MOUNT

The transmission mount should not solidly attach the transmission case to the chassis or a crossmember. The transmission and engine assembly will always be moving about because of torque reactions, so a solidly mounted cast iron or cast aluminum transmission case would readily break because of the flexing.

To adequately support the transmission yet shield it from vibration which could crack the case, some type of rubber mount should be used. All passenger car applications of transmissions employ a rubber-based mount, so the easiest application in a race car chassis is to adapt this rubber mount to a frame crossmember. Most times this can be accomplished by fabricating a bracket off the front of the center frame crossmember, to which the rubber mount can simply be bolted.

Another popular method of mounting the transmission is building a cradle from square tubing which fits beneath the trans. A stock-type rubber mount is sandwiched between the transmission and the cradle, being bolted to both. The square tubing cradle is then bolted to brackets which protrude downward from a roll cage member which runs across the chassis in the driver's cockpit just in front of the seat.

A third method of mounting the transmission is with a metal support bracket which bolts to the top front of the transmission case. This bracket is then lined with a rubber

bushing such as one from a leaf spring eye. A bolt is fed through the spring eye bushing, and is captured on each side of the bushing by a straddle bracket which is welded to the roll cage lateral member in the cockpit.

The first two transmission mounting methods we discussed above are probably equal in their ease of transmission alignment with the bracket when the engine is being reinstalled. The last method described above is

Simplest way to mount the transmission is to weld a bracket to a crossmember, then use the stock application transmission mount to bolt to the bracket.

discouraged because it requires more effort for alignment when reinstalling an engine.

SHOCK ABSORBER MOUNTS

Many differing types of weld-on brackets are used for mounting tie rod end shock absorbers to the chassis. Nothing is more efficient than, and less expensive than, shock eyes made from ends cut off of steering drag links. The ends are cut off the drag links (preferably with a metal band saw) with slightly more than a 1-3/4-inch length of shank protruding below the eye housing. This will allow the shock eye to be installed in 1-3/4-inch tubing all the way through the diameter of it. The shank of the shock eye can be welded both at the top of the tube and at the bottom of the tube. A shock eye installed in this manner will have extreme strength and resistance to bending and breaking because the loads being fed into it by the shock absorber will not stress just one weld. It will spread the load out into the entire tube.

One of the most popular drag links to obtain the shock eyes from is the 1965 to 1970 Ford Galaxy. This drag link has three tapered holes in it of the correct taper and diameter for use with Monroe and Carrera tie rod end shock absorbers. The standard SAE taper for all tie rod ends, ball joint tapers, etc., is .125-inch of diameter change for every linear inch of hole diameter. In other words, if a hole started with a one-inch diameter, this taper would take the diameter of the hole down to .875-inch an inch down the bore from the

Below, these brackets bolt up to a square tubing cradle which holds the transmission.

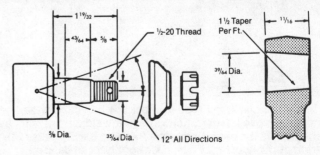

These drawings at right illustrate the requirements for length and taper of a shock absorber mount.

At left is a Ford drag strut shock eye installed in the roll cage tubing. Dotted lines show path of shank through the tube. At right is the type of drag strut piece used to make the shock eye.

beginning. All American automobile tapers have this specification. However, some tapers are drilled deeper than others, so the hole opening diameter on some drag link eyes may be larger than others. The Ford part which we specified above will accommodate the racing shocks. If you have trouble locating this drag link, however, take a racing shock absorber to a wrecking yard with you and fit the eyes to the shock absorber end taper to find one which will fit without remachining.

At left is the machined type of shock mount which is butt-welded to a chassis member. If you use this type of mount, be sure you get good penetration with your weld, or you will have a broken shock mount. Also of note in this photo and the one at the right is the mounting of the Panhard bar. Because this car is equipped with a quick change rear end, the bar mounts in front of the housing. Thus, a straight bar can be used.

Panhard bars are very prone to breakage at their mounting. One cure for this is to triangulate the mount on the housing. This allows a longer connection to be used too so bar does not run at an angle.

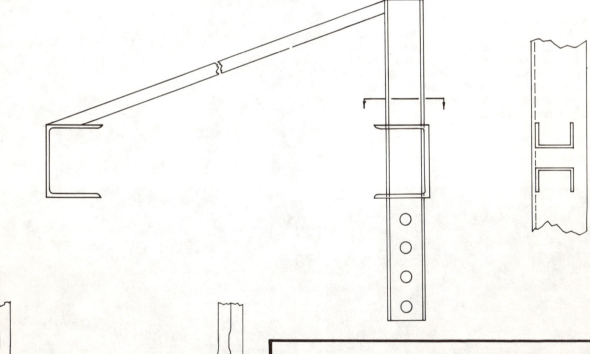

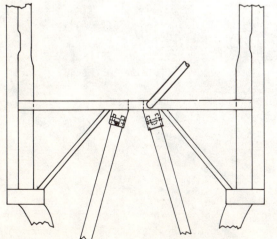

Above left and right, Panhard bar frame mount failure seems fairly common. To combat that, use two pieces of .375-inch thick channel to sandwich the rod end bearing. The mount should extend as far above the frame as below so forces are resisted by a tube tied back to the opposite frame rail. Left, if any trimming of the crossmember is done for driveshaft clearance, a tube should be run from the trimmed section to the upper right side door bar to help the crossmember resist up and down deflections.

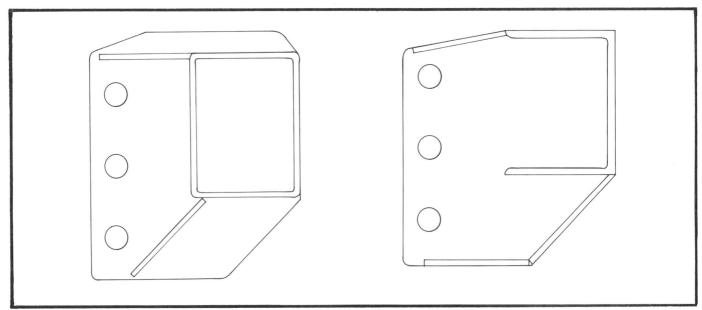

If you want to use a steel plate to build a bracket around the frame rails for the Panhard bar mount, the drawings above will guide you. Use .25-inch thick material for the brackets and .125-inch for the gussets. Use two plates to sandwich the rod end bearing. The photos at right and below illustrate a Watts linkage which pivots on a frame crossmember and attaches to the housing on either side. A good way to go with a quick change.

The Driveshaft

That vital link between the transmission and the differential — the driveshaft — is the last piece of hardware to be fitted and fabricated in the race car. It must be that way because the length of the driveshaft must be measured with the car setting at its normal ready-to-go-racing height and weight. If the length of the driveshaft is measured, or guessed at, with the car in any other condition, the length of the shaft can easily be off.

MEASURING FOR THE DRIVESHAFT

A driveshaft is correctly measured from front yoke center to rear yoke center. To measure the empty space under the car where the driveshaft will reside, first set the car at ride height. Push the transmission slip yoke all the way in, then pull it out ½-inch. Measure — very carefully — from the center of the transmission slip yoke to the center of the companion flange yoke at the rear (in other words, center of the yoke cross to center of the yoke cross). This is the length of shaft you will need. There are two ways that the driveshaft can be obtained.

THE ECONOMY DRIVESHAFT

If you are on a real tight budget and think that you cannot afford to go to a machinist to have a custom driveshaft fabricated, you can go to the local wrecking yard with your trusty tape measure, and start looking for a driveshaft with the exact length which you measured to be required in your car. Remember that the length must be **exact**. If you are

lucky enough to find one the exact length, you must also remember that you need a driveshaft which has a front yoke which will be compatible with the slip yoke of the transmission you are using, and have a rear yoke which is compatible with the rear end companion flange you are using in your car. When you are using a Chevrolet transmission and a Ford 9-inch rear end, it will be hard to find the right front and rear combination. The 1957 to 1970 Chevrolet large passenger car and Chevelle use the same rear universal joint as a Ford 1956 through 1960 with a 9-inch ring gear rear end. Other combinations like this can be worked out by changing the rear U-joint to use other year Chevrolet driveshafts which hook up to the Ford 9-inch ring gear rear end.

Say you find a driveshaft which fits the transmission you are using and it is just slightly longer than the length you need. That's good. Then obtain the yoke you need to go with the Ford companion flange. Take these parts to a machinist, and have him cut the driveshaft to length in a lathe, install the correct rear yoke, reweld, and balance the assembly. Do not try to cut, reweld and balance the driveshaft yourself unless you have a lathe to handle these operations.

When looking for a driveshaft in a wrecking yard, discard all shafts which are two-piece with bonded rubber cushioning, or which are tapered. Inspect the used driveshaft for small dents or rub rings where it has contacted the floorpan. These driveshafts are not good

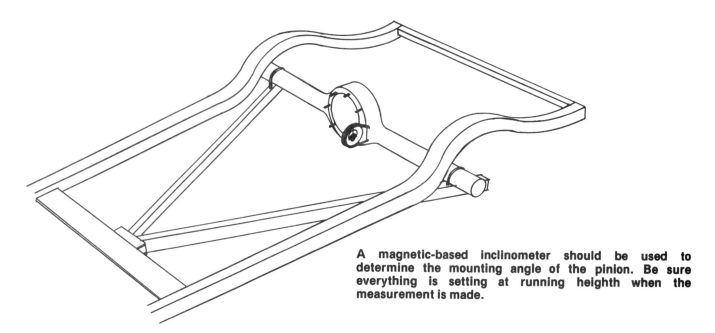

A magnetic-based inclinometer should be used to determine the mounting angle of the pinion. Be sure everything is setting at running heighth when the measurement is made.

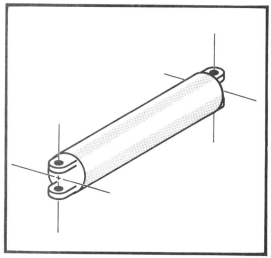

The pivot points of each driveshaft yoke must be absolutely parallel to each other.

A machinist's protractor of this type with a self level is useful in measuring the driveline phasing angles.

prospects for use in your race car because they will be stressed and they will be all but impossible to balance. When you purchase a driveshaft at a wrecking yard, be sure you get the transmission slip yoke, rear yoke, and U-joints along with it. Otherwise you might find the need to go hunting for further parts which usually would tag along for free with the shaft purchase. Even though you have the U-joints with the shaft, discard them and replace them with new Borg Warner Power Brute joints. It is cheap insurance.

THE CUSTOM DRIVESHAFT

To have a custom driveshaft built, take your measurements, and front and rear yokes to a machine shop which specializes in building driveshafts. The shop will furnish the driveshaft tubing cut to length, and will do the welding and balancing. Do you get the idea that a custom driveshaft is not going to cost you much more than the "economy" one? The only difference is the cost of the driveshaft material which you are purchasing, and an extra yoke welded in place. The added cost over the economy method will not be that great, but the added strength and durability will be. This is because the driveshaft material he will select will most likely be 3½-inches O.D., with .095-inch wall. This is substantially sturdier than the stock passenger car shaft tubing, plus you have the assurance it is new material rather than a shaft which may have been nicked, bent, twisted and abused.

BUILDING A CUSTOM DRIVESHAFT

Unless you have a lathe capable of accommodating a driveshaft, you are going to have to find a machine shop which specializes in driveshaft work. But just to give you an

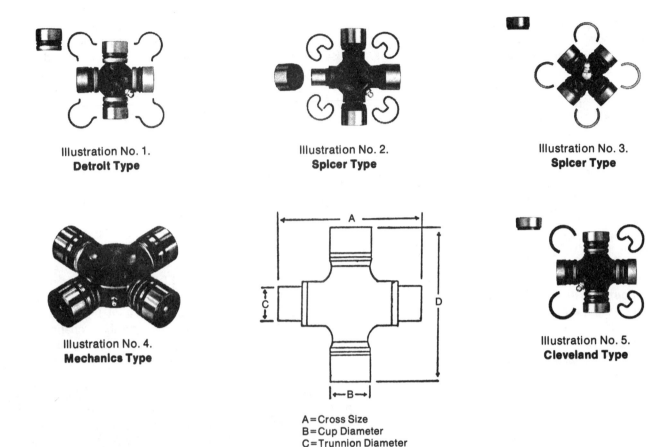

Illustration No. 1.
Detroit Type

Illustration No. 2.
Spicer Type

Illustration No. 3.
Spicer Type

Illustration No. 4.
Mechanics Type

Illustration No. 5.
Cleveland Type

A = Cross Size
B = Cup Diameter
C = Trunnion Diameter
D = Overall Length

These illustrations courtesy of Borg Warner point out the different types of universal joints, along with critical dimensions of the part.

appreciation of what the machinist does, and to give you a background to converse intelligently with him, we present the following information on how the custom driveshaft is built.

Driveshaft tubing, which is generally mild steel D.O.M. tubing finished to more exacting specifications than regular D.O.M. structural tubing, comes in standard lengths of 72 and 108 inches. After the tubing is cut to the required length, its ends are checked for squareness and the inside of the tube is filed to remove any cutting burrs in the material. The yoke for one end is driven in until the shoulder of the yoke contacts the tubing. For this operation, the tubing must be held firm in a vise that will not pinch, nick or compress the tube. The tubing with one end installed is again chucked into the lathe and the yoke and tube are checked for runout — .004 should be a maximum. If the yoke is in the tube square, the yoke is tack-welded to the tubing at four evenly spaced places around the tube. Runout is checked again. Then the yoke is tack-welded again at four more places, and runout is once more checked. If everything is still straight, the welding is finished, then runout is checked again and corrected if necessary. The other end of the shaft is installed while the shaft is still in the lathe, and a level is used on the yoke at each end to be absolutely certain the two yokes are perfectly parallel. This is very critical. If they are not parallel, the vibration will drive you crazy, and likely destroy some components in the car too.

The driveshaft assembly is then balanced. First, though, before we get to that, an explanation is in order about two different balancing methods. The first (and least desirable) method is employed by Detroit carmakers. They spin balance the driveshaft much like tires are dynamically balanced. Where the driveshaft is bowed is indicated as an excessive weight in one direction. To counterbalance that bow or force to one side, a small weight is tacked to the driveshaft in the center of the bow.

The second method, used by driveshaft shops, involves rotating the driveshaft in a lathe or fixture and checking its entire length during rotation with a dial indicator to find any bow. The bow is then corrected rather than counter-weighted.

U-JOINTS

Universal joints take a terrible beating in racing, especially on short tracks where they encounter accelerating forces at least twice a lap. To insure the reliability of these parts, it only makes sense to use the best proven product for the job, which is Borg Warner's Power Brute line of U-joints. These U-joints have a solid cross made from transmission gear steel, have trunions with a finer finish than standard joints, neoprene seals to keep grease in and dirt out, a special heat treatment for greater hardness without brittleness, tapered needles to prevent grooving and flaking, and are quality controlled through constant inspection.

PHASING ANGLES

Forget everything you have heard about how the transmission shaft angle (as seen from the side view) should be parallel to the pinion gear angle. That's for passenger cars which are not designed for acceleration twice a lap. In a race car, the pinion nose lift under acceleration is a much more important design factor. For this reason, the pinion gear centerline is usually placed about 1½ degrees down in coil spring equipped cars, whereas the transmission tailshaft is pointed about 3½ to 4 degrees down in the opposite direction. This trailing arm linkage on coil spring cars limits the amount of pinion nose lift. With leaf springs installed in the rear of a race car, the nose of the rear end will be angled down as much as 5 degrees, depending on the spring rate and the horsepower of the engine. The softer the spring and the greater the horsepower, the more the spring will wrap up during acceleration and move the pinion up. As a guideline for the amount of downward angle, Chrysler recommends 4½ degrees down for its short track Kit Cars, and 2½ degrees down for big track Grand National cars.

The maximum angularity difference between the front and rear U-joints should never add up to more than 7 degrees, in order to minimize horsepower losses and wear,

and eliminate vibrations. This would be figured by measuring the transmission downward angle and the pinion nose downward angle. Say the transmission is angled at 4 degrees, and the pinion down at 1½ degrees. This adds up to 5½ degrees, which is within the 7 degrees of maximum tolerable angularity.

Be sure to measure the angularity on both the transmission and rear end. To do this, use the type of protractor which has a rotating head held in place with a surrounding base, and has a built-in level indicator. Place it on top of the vertical pinion flange at the rear end, and rotate the protractor head until the bubble indicates level. Lock this in place with the set screw and take a reading of the angle. Repeat the same step with the transmission tailshaft flange.

DRIVESHAFT LOOP

What happens when you are driving down the front straightaway and the front universal joint breaks? If you don't have a driveshaft loop to catch the shaft, chances are the shaft will drop to the track and launch your car like a high jumper. To prevent this nasty consequence, build a driveshaft catch loop from 2-inch by ¼-inch steel strap bent into a semi-circular shape. Attach it to the floor pan or preferably the crossmember no more than 6 inches behind the front universal joint. Be sure there is at least 1½ inches of clearance between the driveshaft and loop on each side and on the bottom.

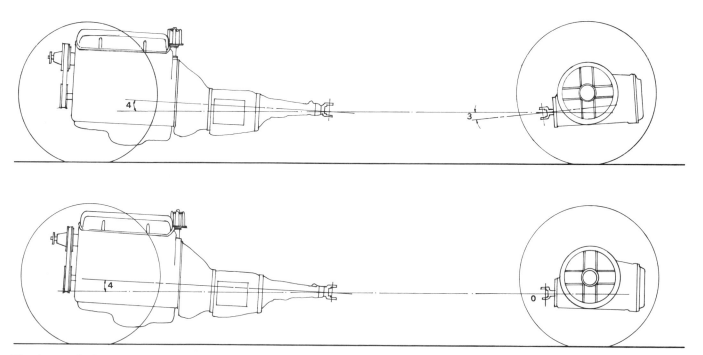

The transmission and rear end phasing angles shown in the top picture should be used for a short track car using leaf springs. A short track car with coil springs would use the driveline phasing shown in the lower picture.

Instruments and Pedals

THE INSTRUMENT PANEL

The first step in laying out an instrument panel is to make a cardboard pattern. Carefully cut a large sheet or sheets of cardboard and fit them into the car's dashboard space. An accurate pattern will enable the builder to see how it will fit, and appear, before a lot of time goes into cutting and building the finished product from 20 gauge steel sheetmetal. The instrument panel should be formed to fit the body sheetmetal, and formed around the roll cage cross brace. The instrument panel **should not** be pop-riveted to the roll cage brace, but rather riveted to tabs welded to the tube.

When the panel is being layed out, all of the gauges to be used should be on hand. This allows the spacing between the gauges to be actually measured. It also gives you the opportunity to be sure of the gauge diameter before the panel is built. Ever run across an advertised 2-inch diameter gauge which is actually 1 7/8-inches in diameter? It sure gives you a problem if the mounting hole has already been cut!

Once the gauge layout and spacing is determined, holes for the gauges should be cut into the pattern, the gauges placed in the pattern, then the pattern with the gauges should be installed so the driver can determine if he has adequate visibility of them with their chosen location with the steering wheel at all positions. If not, it is easy to patch the pattern and shift the locations around.

Because a driver is so busy concentrating on competition during a race, it is difficult at best for him to read a number of gauges and determine the operating condition of his car. To solve this problem and make all the gauges readable at just a glance, all gauges can be mounted so their indicating

A very straightforward, simple dash. Notice the orientation of the gauges — they are mounted so that all needles point straight up when operating conditions are normal.

Use a cardboard template first, then transfer the measurements on the template to the sheetmetal to make this dashboard. Make sure the gauges will not bottom out against the cage tube which runs behind the dash.

needles are in a vertical position when they are showing a normal condition. In this way a driver can quickly spot a problem if a needle is out of position.

If any indicator lights are to be used on the instrument panel, they, along with switches, can be fitted into the pattern in the same manner as we have described above for the gauges. All switches should be positioned where the driver can reach them while strapped in, but not where he could accidentally hit them. A short length of rubber wiper hose can be used as a toggle switch extension so the switches can be reached easier.

Once the panel is constructed and the gauges and switches are installed, they can be pre-wired on the bench — and not while laying on your head on the floorboard. The wires can be neatly bundled this way as well.

You might want to consider using an instrument cluster insert in the dashboard to facilitate troubleshooting of instrument and electrical problems. The instrument insert

can be cut from .063-inch thick aluminum sheet (it is thick to provide stiffness). It is held in place on the 20-gauge sheetmetal dash with t-turn Dzus fasteners.

WHICH GAUGES

At a minimum, all race cars should have these gauges to monitor properly the operating conditions of the engine: tachometer, oil pressure gauge, water temperature gauge, oil temperature gauge and fuel pressure gauge. Only top quality gauges should be chosen, because it does not make sense to monitor the heart of an expensive racing engine with a $1.50 gauge.

The oil pressure gauge should be plumbed with at least a 3/16-inch tubing, with ¼-inch tubing being best. The larger the tubing inside diameter, the more responsive the pressure gauge is. Because oil pressure is so vital in a racing engine, and pressure below a certain minimum can be disastrous to an engine, an auxiliary "idiot light" system should be used

Both these dashboard arrangements use a removeable panel which contains the gauges and switches only. Good idea for quick servicing of a gauge or wiring.

as a supplementary warning to the driver. This can get the driver's attention called to a catastrophic event at a time when he ordinarily would not be observing the gauge. To do this, install a T-fitting at the oil pressure monitoring source and run two lines, one to the oil pressure gauge and one to a warning light. Choose an oil pressure warning light that is set to trigger at a pressure which corresponds to the low point which alarms you, such as 20 PSI.

The fuel pressure gauge is an important item on the panel to help spot potential problems in the fuel pump and fuel filter. The pressure gauge should show at least a steady four pounds per square inch of pressure. If it does not, or the needle fluctuates, go looking for a problem before the problem silences the engine. Every time the fuel pressure gauge line fitting is loosened, be sure to check it for leaks once it is reassembled by starting the engine and subjecting it to normal pressure. A small gasoline leak in the wrong place can mean a nasty fire.

The water temperature gauge is third in importance, and should be treated as such after mounting the tachometer and oil pressure gauge for maximum visibility. The oil temperature and fuel pressure gauges should be mounted on the very outside. All gauges should be plainly labeled in some manner so no confusion could ever result from faulty reading. All temperature gauges should be calibrated in boiling water before installation and use a welding tank regulator to calibrate pressure gauges.

Only a mechanical drive tachometer should be deemed totally accurate and reliable. But even a brand new mechanical drive tach should not be considered accurate until it has been taken to a speedometer repair shop to be calibrated. If the tach is off, it should be on the lower idle side rather than the all important upper range. Use heavy duty drive cable for the tachometer, and route it so it encounters no sharp turns or bends. If a sharp radius is required from the drive coming out of the distributor, obtain an angle drive fitting from a speedometer shop to make the bend.

THROTTLE LINKAGE

The end result you are looking for in a throttle linkage is a clean, simple linkage which is as trouble-free as possible, easy to hook up, and puts as small a hole as possible in the firewall.

The location of the pedal and bellcrank assembly on the firewall should be determined by finding a comfortable position for the driver's foot. Locate the pedal attachment point on the firewall by sitting the driver in the driving position and drawing his right foot to a position where it feels at ease. Locate the bracket assembly and attach it to the firewall. The linkage or cable can then be run to the carburetor, in as straight a line as possible.

Keep the foot pedal travel long enough to enable the driver to apply power smoothly. Two adjustments which dictate this is the length of the arms on the bellcrank (short arms means fast travel and little driver modulation control), and how far from the centerline of the crank on the carburetor the linkage arm or cable is connected.

One of the best simple and neat throttle bellcrank assemblies to use with a mechanical linkage is from a 1961

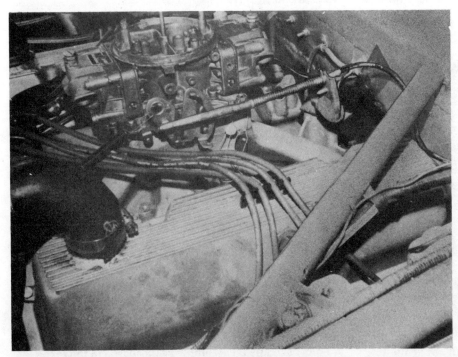
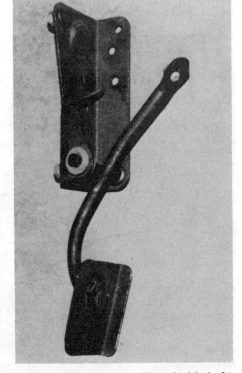

Engine at the left uses a cable throttle linkage acquired from a Ford product such as a Galaxy, Cougar or Thunderbird. At right is the Chrysler pedal and bracket assembly part number 3496383.

This throttle linkage bellcrank is from a 1961-63 Buick Special. Neat little unit.

Use a minimum of two throttle return springs on the carburetor. Be sure they mount in different places on the carb, and have a sturdy anchor point on the other end. Springs should hook to carb, not linkage.

to 1963 Buick Special. See the accompanying photo for more details. Another one is found in a 1965 and 1966 Ford. If you search under the dashboard of several passenger cars, you may spot other simple and light bellcrank assemblies which can work as well, such as are illustrated in accompanying photos.

When using a mechanical throttle linkage, be sure items like the air cleaner, spark plug wiring and tachometer drive cable cannot become entangled in the linkage, or hinder its free movement. Above all else, use **two** quality return springs on the linkage to have a safety return in case one should fail.

Sometimes a flex throttle cable is preferred, but be sure the one you choose has no rubber or plastic bushings in the cable housing or brackets. For a simple and lightweight bellcrank and mounting bracket assembly for a flex cable throttle linkage, use Chrysler Corporation part number 3671723 or part number 3496383. A flex cable assembly to use with the 3671723 bracket is Chrysler part number 3577525. Another high quality cable is Morse Instruments' "Red Jacket" line, found at boat dealers and marine hardware stores. For rigidity, fabricate your own cable locating bracket from 1/8-inch thick steel plate. Be sure the flex cable is straight and not kinked, and be sure there is no possible way anything could ever interfere with its operation. Check the cable weekly for signs of wear and fraying. Never use motorcycle throttle cables for a race car, and note that heat can cause cable-to-housing bind in some types of cable.

Notice how the brake and clutch pedals both set at the left of the steering wheel, yet are positioned comfortably for a driver's foot.

A Ford-based pedal assembly as discussed in the text.

BRAKE PEDAL

The brake pedal should be located on the left side of the steering column if the driver is going to brake with his left foot, and on the right side of the steering column if he is going to brake with his right foot. This will give the driver's braking foot a straighter shot at the pedal for better leverage, and eliminates his leg from crossing under the steering column which could cause a broken leg in an impact.

The brake and clutch pedals can be housed in a standard pedal support housing obtained from most any passenger car equipped with a manual transmission. Look for a housing which is simple and lightweight, such as early Camaro or Maverick.

Do not rely on stock rubber pedal covers to stay glued onto the pedals. Secure them with pop rivets.

CLUTCH LINKAGE

A racing clutch uses high spring pressure, so stock passenger car clutch linkages are usually not stiff enough to handle the job. If the linkage deflects during pedal application, the clutch probably will not release completely. To combat this, fabricate a cross shaft from heavy wall tubing, using the stock piece as a pattern. The actuating rod should be constructed from at least 3/8-inch diameter solid steel rod, and have adjustable rod end bearings fitted on each end.

A very sensible and practical means of actuating the clutch is by means of a hydraulic cylinder and slave cylinder. It weighs less than mechanical linkage, is less prone to binding because of engine movement, and it makes engine changing easier (take one bolt off and the linkage is free). Be aware that clutch actuation requires the displacement of more fluid from the clutch master cylinder than does the operation of a braking system. To prevent this from slowing down your clutch operation, use at least 3/16-inch I.D. fluid line. Make sure the line is routed away from the headers, and that it is in a position where it cannot get kinked or crimped (especially when changing engines).

Another alternative to the mechanical clutch linkage is a tension cable in a flexible housing. The Mustang II, Pinto and Capri all use this type of clutch actuation, and the pedel, cable and bracket can be readily adapted to a race car (see diagram of Mustang II arrangement).

PEDAL STOPS

Everyone realizes a driver uses his feet and legs to their utmost capacity in the heat of a race, so to prevent him from bending linkages, stops MUST be used on the throttle, brake and clutch pedals.

An adjustable stop for the throttle pedal can be built by welding a nut to the floorboard or firewall and threading a

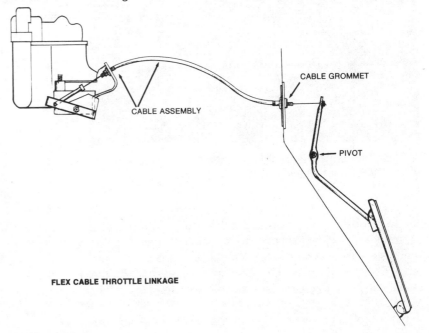

FLEX CABLE THROTTLE LINKAGE

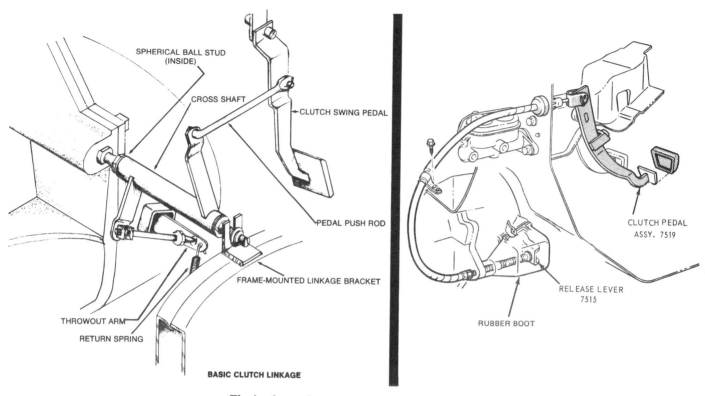

SPHERICAL BALL STUD
(INSIDE)

CROSS SHAFT

CLUTCH SWING PEDAL

PEDAL PUSH ROD

FRAME-MOUNTED LINKAGE BRACKET

THROWOUT ARM

RETURN SPRING

BASIC CLUTCH LINKAGE

CLUTCH PEDAL
ASSY. 7519

RELEASE LEVER
7515

RUBBER BOOT

The basic mechanical clutch linkage is shown at the left. At right is the Mustang II flex cable clutch linkage.

A hydraulic clutch is simple to mount to dismount, and offers many positive benefits for a race car.

bolt in it. Position the stop under the pedal, not against a bellcrank rod. Adjust the stop so it contacts the pedal just as soon as the throttle plates in the carburetor are wide open, otherwise the driver will be putting an extreme strain on the throttle shaft of the carburetor (and it will break off rather easily).

The clutch stop prevents a driver's leg from generating

enough force on the linkage that could break something, plus it keeps a diaphragm clutch from going over center. Position the stop as close to the pedal itself as possible, on the floorboard or firewall. Also install one against the shaft extending from the dash so one inch of free play can be adjusted (so throwout bearing does not continually spin). Be sure to use a return spring on the clutch pedal shaft.

The brake pedal stop is designed to prevent linkage or master cylinder damage once the master cylinder piston has bottomed. Build the adjustable stop to contact the pedal shaft from the firewall. With disc brakes, you must allow the brake pedal push rod to have an end play of .015 to .020-inch so that the master cylinder piston can return to position to uncover the fluid return port. It is also important to use a brake pedal return spring so the master cylinder piston will return to its neutral position. If these steps are not taken, the lines will pump up with each succeeding brake application, and cause the brakes to lock up, and then they will be removing your car from the track with two wreckers. With a drum braking system there are return springs inside of each wheel cylinder to take care of this problem.

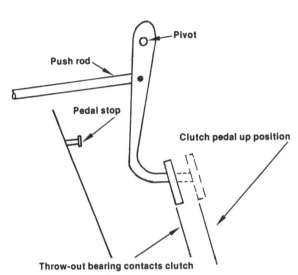

An aluminum pillow block to hold steering shaft.

One inch of free play should be adjusted in the clutch linkage.

STEERING COLUMN

Support the steering shaft with a 3/4-inch rod end bearing bolted to a bracket welded to the roll cage dash cross tube. At least one, and better yet two, universal joints should be contained in the steering shaft to prevent a head-on crash making the steering shaft a spear to the driver's chest. The steering wheel should be of a dished design, and be lined with a **rigid** foam in its center.

This steering column [at right] is adjustable for wheel heighth to fit driver's preferences.

The rod end bearing should be mounted on the steering shaft as close to the wheel as possible. Also affix a 3/4-inch set collar on each side of the bearing so that in case of a crash it will help resist the shaft from spearing the driver.

The Body, Body Supports and Cockpit

THE BODY SYSTEM

Proper body design and construction for a race car has three functions — provide a proper and pleasant work area for the driver (as well as providing ease of internal maintenance), provide a workable aerodynamic shape along with reduction of drag, and contribute to the total stiffness and strength of the completed structure.

STRIPPING THE BODY

Whether your source for a body is a new car, a used car, or the more traditional source of a body from a wrecking yard, the body sheetmetal **must** be stripped of all extra metal, to reduce the body weight as much as possible. Cut body structures such as roof inner panels, door and quarterpanel inner structure, wheel wells and fender panels, trunk hinge supports, rocker panels (on a non-unibody car), and all engine compartment sheetmetal support structures. If it is allowed by the rules of your racing association, remove the stock firewall between the engine compartment and driver cockpit, and replace it with one made from 20 gauge steel sheet stock. The stripped-out body should be little more than just the sheet metal exterior skin.

If you are planning to build a fairly late model car such as a Nova or Chevelle that still has sheetmetal parts serviced by the manufacturer through dealers, you can purchase all the exterior sheetmetal, such as door skins, roof, quarterpanels, etc., and weld them together to make a body.

You can help support the inner edges of the fenders, hood, deck lid, etc., by welding ½'' x ½'' steel square tubing to them as a liner.

A popular body material today which is replacing the steel sheetmetal on race cars is fiberglass. There are several companies in the country (most notably McGrath Fiberglass in El Cajon, Calif., and Speedway Motors in Lincoln, Nebraska) which will supply almost the entire body in fiberglass for the more popular cars such as a Nova and Camaro. There are fiberglass front fender, hood and grille

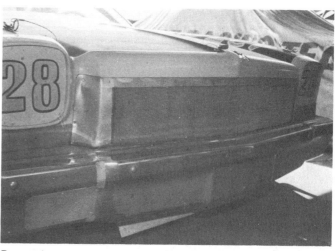

Proper body and exterior sheetmetal design comes together to provide a workable aerodynamic shape.

Short track aerodynamics can be improved if air is prevented from getting underneath the car. A piece of conveyor belting material can get the job done nicely, riveted to the bumper. It is also flexible so it takes a lot of abuse.

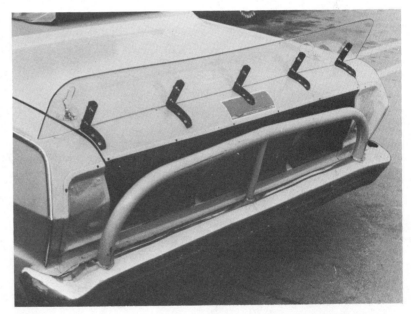

This huge spoiler is a help on a short track, but does not detract from stock appearance of car because it is made from Lexan.

nose pieces, doors, tops and trunk lids available for Novas and Camaros. The fiberglass reproductions of the sheetmetal pieces are excellent in reproduction, are more durable than sheetmetal, and cost less than sheetmetal. Be sure when you purchase fiberglass body parts for a race car that the parts are made by the hand laminated method and not the chopper gun method. The parts made with a chopper gun are less costly, but are vastly inferior to hand laminated parts in quality, strength and thickness uniformity.

Many people look to a method called acid dipping to lighten their body sheetmetal. Acid dipping was used extensively by the factory racing teams running the Trans Am racing series in the late 1960's, and is used extensively in drag racing as a weight reducing method. But for the rigors

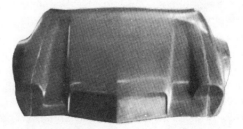

Several manufacturers are supplying fiberglass body panels for race cars. They are amazingly durable.

The body should be stripped to the bare minimum, as above, then set on the chassis to fit cage and sheetmetal.

of modern short track "contact" stock car racing, acid dipping the body is not practical. Acid-dipped sheetmetal will behave like heavy gauge aluminum foil, making it quite susceptible to damage, and difficult to repair.

INTERIOR SHEETMETAL

Fitting, cutting and installing the interior sheetmetal in a race car can be a craftman's delight — or nightmare — depending on how the situation is approached. The easiest way to fabricate it is to make cardboard templates of each piece, such as front firewall, rear firewall, etc. Use light cardboard, pencil, ruler, scissors and tape, and build your interior first from the cardboard. Make sure each piece fits together with its adjoining pieces. Then, after you are assured of the fit, make the panels from 20 gauge steel sheet material. Some people, in the interest of saving weight, are switching to 24 gauge sheet steel for the floor pan, fender wells and firewalls. This can be done if the car has a rigid chassis. If 24 gauge is used with a somewhat flexible chassis, the sheetmetal will look like a road map after six races. The other difference is in welding. The difference between welding 20 gauge and 24 gauge is like the difference between 3/16-inch thick steel plate and paper. A wire welder is definitely required to weld these materials. With 20 gauge material, .035-inch wire can be used for welding, but 24 gauge requires the use of .030-inch wire.

Some racing associations require that the floorpan be a stock appearing pan, rather than being fabricated from sheetmetal. In that case, use a floorpan from a 1967 Ford Galaxy, 1971 Ford Torino, or 1968 Camaro. These floorpans are relatively flat with little or no depression in them for footwells. The floorpan should be welded to the frame rails at each side of the chassis, and is supported by and tacked to the frame crossmember in the middle.

Use a cardboard template to help you fashion any interior sheetmetal.

To keep the interior sheetmetal appearance professional, careful measurements should be made where roll cage tubes will protrude through the sheetmetal. This helps the builder limit the size of the hole he cuts right to the precise size required, thus preventing oversized holes and unsightly sheetmetal patches. Cutting cardboard templates can help this fitting procedure. By all means, fit the interior sheetmetal to the chassis before the finish welding is done on the roll cage.

For another professional beautifying touch with the interior sheetmetal, add beading to the panels. The beading also significantly adds to the strength of the sheetmetal.

SHEET GAUGES

Gauge No.	STEEL SHEETS		GALVANIZED SHEETS		STAINLESS STEEL SHEETS			ALUMINUM SHEETS	
					Wt., Lbs. per Sq. Ft.				
	Weight Lbs. per Square Foot	Thickness in Inches	Weight Lbs. per Square Foot	Thickness in Inches	Straight Chromium (400 Series)	Chromium Nickel (300 Series)	Approx. Thickness in Inches	Weight Lbs. per Sq. Ft. (1100)	Thickness in Inches
38	.25000	.0060						.0558	.00396
37	.26562	.0064						.0627	.00445
36	.28125	.0067						.0705	.00500
35	.31250	.0075						.0791	.00561
34	.34375	.0082						.0888	.00630
33	.37500	.0090						.0998	.00708
32	.40625	.0097	.56250	.0134	.3708	.3780	.010	.1121	.00795
31	.43750	.0105	.59375	.0142	.4506	.4594	.011	.1259	.00893
30	.50000	.0120	.65625	.0157	.5150	.5250	.013	.1410	.0100
29	.56250	.0135	.71875	.0172	.5794	.5906	.014	.1593	.0113
28	.62500	.0149	.78125	.0187	.6438	.6562	.016	.1777	.0126
27	.68750	.0164	.84375	.0202	.7081	.7218	.017	.2002	.0142
26	.75000	.0179	.90625	.0217	.7725	.7875	.019	.2242	.0159
25	.87500	.0209	1.03125	.0247	.9013	.9187	.022	.2524	.0179
24	1.0000	.0239	1.15625	.0276	1.0300	1.0500	.025	.2834	.0201
23	1.1250	.0269	1.28125	.0306	1.1587	1.1813	.028	.3187	.0226
22	1.2500	.0299	1.40625	.0336	1.2875	1.3125	.031	.3567	.0253
21	1.3750	.0329	1.53125	.0366	1.4160	1.4437	.034	.4019	.0285
20	1.5000	.0359	1.65625	.0396	1.5450	1.5750	.038	.4512	.0320
19	1.7500	.0418	1.90625	.0456	1.8025	1.8375	.044	.5062	.0359
18	2.0000	.0478	2.15625	.0516	2.0600	2.1000	.050	.5682	.0403
17	2.2500	.0538	2.40625	.0575	2.3175	2.3625	.056	.6387	.0453
16	2.5000	.0598	2.65625	.0635	2.5750	2.6250	.063	.7163	.0508
15	2.8125	.0673	2.96875	.0710	2.8968	2.9531	.070	.8051	.0571
14	3.1250	.0747	3.28125	.0785	3.2187	3.2812	.078	.9038	.0641
13	3.7500	.0897	3.90625	.0934	3.8625	3.9375	.094	1.015	.0720
12	4.3750	.1046	4.53125	.1084	4.5063	4.5937	.109	1.139	.0808
11	5.0000	.1196	5.15625	.1233	5.1500	5.2500	.125	1.279	.0907
10	5.6250	.1345	5.78125	.1382	5.7937	5.9062	.141	1.437	.1019
9	6.2500	.1495	6.40625	.1532	6.4375	6.5625	.156	1.613	.1144
8	6.8750	.1644	7.03125	.1681	7.0813	7.2187	.172	1.812	.1285
7	7.5000	.1793						2.035	.1443
6	8.1250	.1943						2.284	.1620
5	8.7500	.2092						2.565	.1819
4	9.3750	.2242						2.881	.2043
3	10.0000	.2391						3.235	.2294

A beautiful interior sheetmetal job. Beading in the metal adds greatly to the strength.

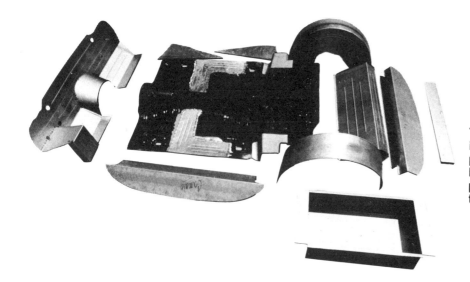

The complete interior sheetmetal before installation. Using the cardboard templates will help you preserve the good looks of your car by pinpointing the exact position and size of all holes for roll cage tube members.

This is the Ford floorpan we referred to in the text.

The sheetmetal and the basic roll cage get fitted together before anything is finish-welded in place.

The telescoping hood suppord brackets shown in these three photos come from most passenger cars with a hatchback rear, such as the Datsun 280-Z, Ford Mustang II with hatchback or Chevy Nova with hatchback.

BUMPERS

There are probably as many ideas about bumpers and bumper bracing as there are race cars in existance. Our idea for a bumper is to first enhance the appearance of the race car and not detract from it, second to protect the front structure of the sheetmetal and frame, and third to achieve these other goals with a minimum of expense. Another consideration for bumper choice can be the aerodynamics the bumper provides.

The best bumper to use, if they are available for your car or will fit your car, is the late model aluminum bumpers. These bumpers are available on late Novas, late Camaros and late Chevelles. You can also find some good swaps, such

as a '76 Camaro bumper for a '69 Chevelle. The aluminum bumpers are lightweight, but very hard to destroy. We have seen some cars take a hard enough lick to destroy sheetmetal, but the aluminum bumper escaped unscathed. A good economy tip: the aluminum bumpers cannot be rechromed, so check body shops to pick up good bumpers with just minor scrapes in the chrome.

BUMPER BRACING

The days of bouncing hard off the first turn wall and still going on to win the main event are behind us. Thus, the good old days of using 120 pounds of bumper bracing and armor plating are also behind us.

Bumper bracing at left and below right adequately protects tires and front end components, and allows bumper to take quite a blow without coming loose. A much more solid bumper bracing is desirable for the rear end of the car, below left.

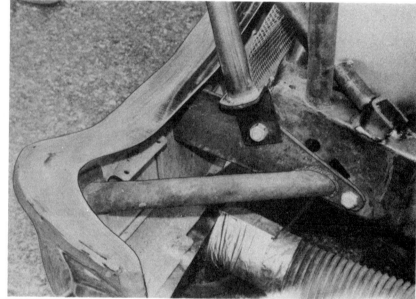

There are two objectives in designing bumper bracing. First, it should be designed to withstand normal pushing, shoving and tapping without loosening the bumper, yet be weak enough to withstand harder forces to cushion the main frame structure from damage. The second objective is to have sufficient side bracing on the bumper ends to protect the tires.

A good bumper mounting design starts with two bumper brackets, extended from the bumper to the frame horns, made from flame-cut ¼-inch thick steel. Each bracket should mount to the bumper with at least two Grade 5 half-inch bolts. The flame-cut brackets should mount to the frame horns with two bolts, one in front of the other as seen from the side view. The front bolt should be a 3/8-inch ungraded low carbon steel bolt. The second bolt should be at least Grade 5, ½-inch. The idea of this type of mount is that the front low grade steel bolt will be sheared with a moderate impact, absorbing most of the impact energy. The Grade 5 bolt will remain intact and allow the bumper to pivot upward to dissipate force instead of bending any components. This will allow a car to take a moderately hard bump without damaging front components or knocking the front bumper loose. If the car takes an impact hard enough to shear the Grade 5 ½-inch bolts, be ready for some instant body work because there is nothing that would have prevented damage anyway.

Above and right, two ideas for supporting front fenders. The square tubing at right will take more impact without transferring damage to the chassis than round tubing.

Left, front collapsible structure is made of square tubing. Note how hood pin retaining bracket is supported. Below, window retaining clips are made of 3-inch by 1-inch by 1/8-inch material. Use 3 clips above windshield, 2 below. The edge of the windshield glass should be lined with gray tape to prevent stress cracks from developing and spreading.

Another idea for mounting the late model aluminum bumpers is to use their stock leaf-type spring to mount them to the frame. Mount the springs to L-shaped brackets welded vertically to the side of the frame. This allows the brackets to be torn away from the frame before the frame is damaged, should the bumper take a very hard impact.

PAINTING THE RACE CAR

There are four types of paint to choose from for a race car — regular enamel, acrylic enamel, lacquer and DuPont's Imron.

For a show car, lacquer is the best paint to use. It will have great depth and gloss. But for a race car, something more

Start your painting at the front of the car and paint each side and the top front, front to rear. Note that the floor has been watered down to prevent static electricity, to settle dust, and to carry away overspray.

practical and useful is required. Lacquer is hard to spray, it needs many coats to adequately cover, and it must be rubbed, waxed and buffed to achieve a good gloss.

Regular enamel has the advantage of producing a gloss without the need of hand rubbing and waxing. But it has a major drawback of wrinkling when you paint over it if it has not been applied for at least six months. A race car surely will have to be retouched more often than that!

For the economy budget, acrylic enamel is the answer. Its preparation does not require anything special, and it has a good self-gloss. No rubbing compound is needed. And it has the advantage of drying in about one hour (compared to as much as six to twelve hours for the old style enamel), and the drying time can be accelerated with the addition of a catalyst (such as Ditzler's Delthane) which also adds some chip resistance to the finish. What this means to the racer is that he can repaint the rear quarterpanels of his car and still reinstall the rear end in one night without the worry of messing up the paint. Acrylic enamel has a further advantage of being able to be spot-painted on a fender or a door with no problem of matching or blending.

DU PONT'S IMRON

Imron is a new polyurethane air-dry enamel developed and sold by DuPont. Originally developed for commercial aircraft which are subjected to large temperature extremes as well as impact and abrasion abuse, Imron is now available in over 1,000 colors as a fleet truck finish. What this means for race cars is a finish which offers superior gloss retention, and maximum chemical resistance, adhesion, corrosion resistance, impact resistance and cleanability.

Experienced painters claim Imron is a very easy finish to apply. It can be sprayed like lacquer or enamel. With normal air drying it is completely dried in six hours, and with an accelerator added to the paint, it can dry in as short a time as two hours.

The Imron paint is a two part formula. You purchase a gallon of the paint which actually has three quarts of paint in the can, then add a quart of special activator.

There are drawbacks to Imron however. The cost of the paint and activator are quite a bit higher than that of acrylic enamel (although some people report tht Imron covers completely with less paint than acrylic enamel). There are also reports that it radiates a paint odor forever when subjected to high heat such as radiated from exhaust pipes.

WHICH COLORS

Everybody has his own color preference for the exterior of a car, but consider this: lighter colors tend to be much more visible at night. A light colored car will stand out much more readily during night racing and be more attractive to the fans in the stands. (Maybe you can paint your car completely navy blue and be anonymous to the other racing competitors!)

For the interior and under-the-hood areas of a race car, we strongly recommend dark dray shades ranging anywhere from medium gray to black. Whites and light grays may look nice when first painted, but they **quickly** show the signs of grease, oil and dirt, and distract from the overall appearance of the car.

A BUDGET PAINT JOB

An attractive and durable paint finish need not be costly. With a little work and effort, you can have the best appearance for a cost affordable to any budget.

First of all, any professional painter will tell you that a good paint job is all in the preparation. By doing all of the preparation yourself, you can save considerable money. Start the preparation with the finished chassis before the body is added. Clean it thoroughly with a metal prep, then spray it with a non-sanding primer. This will keep parts from rusting that you will not be able to reach with a spray gun after the car is completed. Be sure to seal off all areas you don't want embedded with paint such as threads.

If you are unfamiliar with the proper primer to use as well as priming and sanding techniques, talk to a salesman at an auto paint store. These people are knowledgeable and usually very willing to give help and advice. Another source of information is from experienced racers who have faced several seasons of painting and repainting their race cars.

Once the car is properly prepared, you can spray it yourself, have a friend spray it, or take the car to a professional painting company such as Earl Scheib, which will paint an entire car for under $40. If you paint the car yourself, be sure to do it in a clean area protected from winds and breezes. As a further measure to hold down dust and static electricity and carry away overspray, wet down the floor around the car before spraying it. Be careful about proper ventilation in the painting area, and do not spray or leave open cans of paint or thinner near any open flame.

When you paint the car yourself, start at the front and work toward the back. For example, work on the same line around the front fenders and hood. Do not start at the front fender on one side and work all around the car back up to the front fender on the other side. Doing it in this manner will allow one edge of the paint to set-up and get frosty. When the fresh paint is painted over this edge, the frosty appearance will remain. With acrylic enamel, two or three coats should cover the car entirely. Allow about 15 minutes of drying time between coats, assuming a 70-degree air temperature and moderate humidity in the air. One gallon of acrylic enamel should cover the entire exterior of a car, with another gallon required for the interior.

Buying Wrecking Yard Parts

For a great deal of the major hardware required to construct a race car, the wrecking yard is going to be a good sound parts source. To make an economical buy, and to spot the correct pieces you need, we have provided the following information to guide you. We will show you how to spot the high performance or heavy duty pieces most desirable for use in a race car. And, we will show you how to inspect the condition of used parts in order to know what you are buying.

IDENTIFYING FORD ENGINES

Ford provides very detailed product identification tags on all of their major hardware, such as engines, transmissions and rear ends. On Ford engines, the tag is usually mounted at the front of the engine or under the coil. It tells the displacement and model year of the engine. The information can also help spot specially equipped blocks such as the Cleveland Cobra Jet and H.O. designations which mean four-bolt main caps. If for some reason the numbers on the tag are obliterated, or the tag is missing, supply your Ford dealer with the manifold casting number or distributor and carburetor numbers. A good parts counterman will be able to interpret for you. He can also interpret the codes on the tag for you. If he cannot, the definitions of the codes are in service manuals, and the Hollander Interchange Manual.

Boss 302 engines can be identified by a part number cast on the block. The part numbers beginning C9ZE or D0ZE identify a Boss 302. You can also spot one by its canted

If you know what you are looking for and how to identify the good parts, there can be gold in piles like these for you.

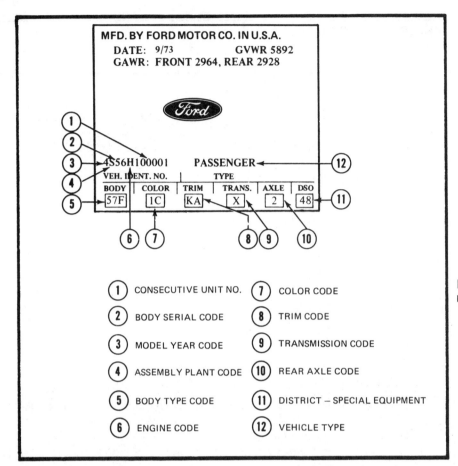

MFD. BY FORD MOTOR CO. IN U.S.A.
DATE: 9/73   GVWR 5892
GAWR: FRONT 2964, REAR 2928

4S56H100001   PASSENGER
VEH. IDENT. NO.   TYPE

BODY	COLOR	TRIM	TRANS.	AXLE	DSO
57F	1C	KA	X	2	48

(1) CONSECUTIVE UNIT NO.   (7) COLOR CODE

(2) BODY SERIAL CODE   (8) TRIM CODE

(3) MODEL YEAR CODE   (9) TRANSMISSION CODE

(4) ASSEMBLY PLANT CODE   (10) REAR AXLE CODE

(5) BODY TYPE CODE   (11) DISTRICT – SPECIAL EQUIPMENT

(6) ENGINE CODE   (12) VEHICLE TYPE

Ford product vehicle identification numbers and interpretation.

valve heads and chrome or cast aluminum valve covers (assuming the valve covers are original equipment).

The Ford 351-Cleveland engine, which is more desirable for high performance than the 351-Windsor engine, can be spotted by its canted valve heads. The 351-C also has very wide valve covers. Beware that the potential 351-C you have spotted is not a 400 cubic inch version of the Cleveland small block. It is identical in outward appearance. The coded tag can help you sort out the right part. There are three positive identifications for the 351-C four barrel Cobra Jet engine, which has four bolt main caps and large cylinder head ports: 1) the four-barrel Autolite carburetor has a spread-bore pattern, 2) the oil dipstick is calibrated for five quarts, and 3) the distributor is a dual-point dual diaphragm unit. 351-C two-barrel versions have been built, but the head ports are much smaller. Any 351-C with a mechanical camshaft also identifies a desirable engine for you.

The 428 cubic inch Ford engines with high performance goodies inside of them have the designation "C-8" cast on the cylinder heads between the center two exhaust ports. Ford 427 high performance engines (if there are any to be found any more) have the designation "C-5" cast on the cylinder heads between the center two exhaust ports.

IDENTIFYING CHEVROLET ENGINES

First of all, the most obvious differences in Chevy small

blocks is between the four-inch and 4 1/8-inch siamese bore blocks. These can readily be spotted by counting the freeze plugs on the engines. The 400 cubic inch siamese bore block has three freeze plugs on each side while all other small blocks have only two. To spot the high performance four-bolt main block, look for one which has an extra oil pipe plug on the left side above the oil filter pad. To double check, then, pull the pan.

To pick out the desirable big block Chevy engines, look for the words "HI-PERF" cast on the cylinder heads inside the rocker covers. Aluminum heads residing on a big block would also tell you it is a high performance engine with four-bolt mains.

There is a machined pad on the block in front of the right hand cylinder head on all Chevrolet V-8 engines on which a number and letter code is stamped. This code can be used to identify the more desirable high performance engines Chevy has produced. The table included in this chapter lists most of the high performance codes. The pad has two numbers on it. The sequence of numbers such as 61026CTC would be the one you want for identification — the CTC at the end tells you it is a 350 cubic inch block with a factory horsepower rating of 360, which would be a high performance block. The second series of numbers on the pad tells you the model year the engine was manufactured. For example, in a number 13T211398, the second digit (3)

CHEVROLET ENGINE CODES		
Cubic Inches	Horsepower	Codes
302	290	DZ, MO, MP
327	300	H, RB, SB, ZB, ZL
327	325	EP, ER, ES, ML
327	340	RE, RD, SD
327	350	EC, HD, HK, HP, HO, HT, HU, HV, HW, ZG, ZI
327	365	HH, HL, RT, RE, RF, RR, RV, RX
350	300	CNJ, CNK, CNO, CNQ, CNR, CNS, CNT, CRE, HA, HB, HE, HG, HH, HK, HN, HY, HZ
350	350	CTN, HW, HX
350	360	CTB, CTC
350	370	CTR
396	375	BG, EG, EX, JD, JH, JJ, JL, JM, KC, KD, KE, KF, KG, KH, KI, MQ, MR, MT
396	425	IE, IF
402	375	CTH, CJJ, CJL, CJM, CKC, CKE, CKL, CKO, CKP, CKQ, CKT, CKU, CTY
427	400	IM, FO
427	425	ID, IP, LD, LS, MD
427	430	LO, LV
427	435	IR, IT, IU, JA, JE, LP, LR, LT, LU, LW, LX
454	450	CRV

stands for 1973.

MOPAR ENGINES

To identify an engine, look for the engine number cast on the cylinder block, either below the cylinder head or on the pan rail on the left hand side. The number is usually 14 digits long. The third, fourth and fifth numbers are the cubic inch displacement of the engine. If you find an eight-digit number stamped on the pan rail of the block behind the right engine mount, it is the vehicle identification number. The first character is the model year. The most desirable 340 block for high performance use is the T/A 340, but it is very rare. Should you ever run across one, you can readily spot it because it has a number cast on the left side of the block, above the freeze plugs, which contains the characters "T/A 340."

TRANSMISSIONS

The accompanying picture chart (reproduced courtesy of Hurst) will help you identify almost any transmission visually. In addition to the picture identifications, we add the following information which can help you identify various transmissions.

FORD T&C TOPLOADERS

These strong performance transmissions from Ford are referred to as "top loaders" because their covers are located on top of the transmission instead of on the side. It is also referred to as the T&C transmission, after the designers, Thompson and Collins.

All Ford transmissions have a stamped metal tag attached to them, usually captured by one of the bolt holes in the cover. The top row of characters will be the product identification code. All four speed T&C toploaders start with the product code "RUG."

There are four different versions of the Ford toploader: (1) large input shaft with short tailhousing, (2) large input shaft with long tailhousing, (3) small input shaft with short tailhousing, and (4) small input shaft with long tailhousing. The large input shaft is 1 3/8-inches in diameter. The small input shaft is 1 1/16-inches in diameter.

Look for the Ford toploaders in any Ford products built since 1965 which have a large displacement engine (390, 427, 428, 429), or high performance small blocks.

There are two ratio versions of the toploader, wide ratio and close ratio. The close ratio trannies have a 2.32 first gear ratio whereas the wide ratio model has a 2.78 first gear. Be sure to check the first gear ratio to determine which model you have picked up.

MUNCIE TRANSMISSIONS

Both Saginaw and Muncie transmissions have been used interchangeably through all GM divisions (except Cadillac of course). The Muncie is the only version to use for racing. There are three varieties of the Muncie: (1) the wide ratio M-21, (2) the close ratio M-21 and the M-22. The M-22, which is not manufactured any more, is the most desirable. The three transmissions can be identified from each other by grooves around the input shaft. The M-22 does not have any grooves. The close ratio M-21 has one groove, and the wide ratio M-21 has two grooves.

MOPAR TRANSMISSIONS

There are two versions of MoPar four speeds. The type

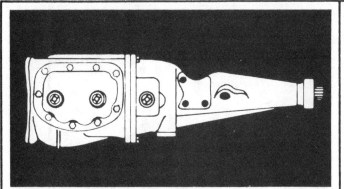

411 Borg-Warner T-10, Super T-10
9 bolt curved bottom side cover

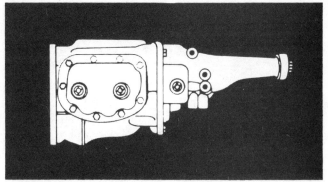

412 Borg-Warner T-10, Super T-10
9 bolt curved bottom side cover

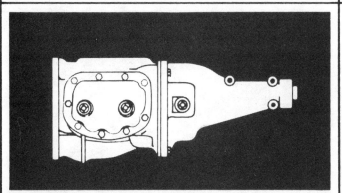

413 Borg-Warner T-10, Super T-10
9 bolt curved bottom side cover

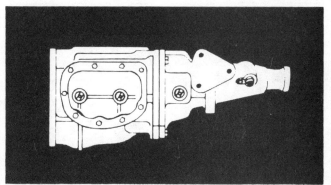

414 Borg-Warner T-10
9 bolt curved bottom side cover

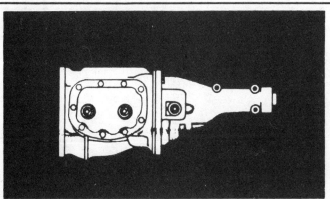

415 Borg-Warner Super T-10
9 bolt curved bottom side cover

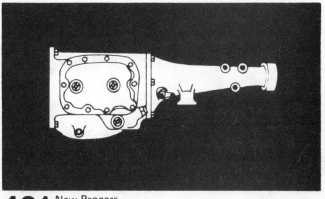

421 New Process
10 bolt side cover

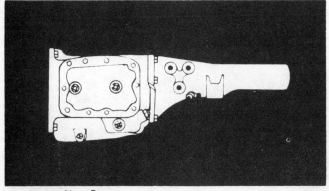

422 New Process
10 bolt side cover

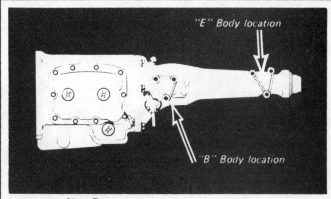

423 New Process
10 bolt side cover

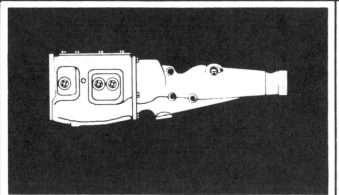

431 Ford T & C
10 bolt top cover

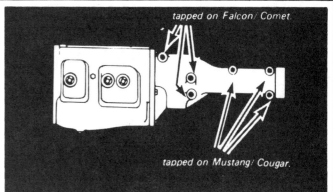

tapped on Falcon/Comet.

tapped on Mustang/Cougar.

432 Ford T & C
10 bolt top cover

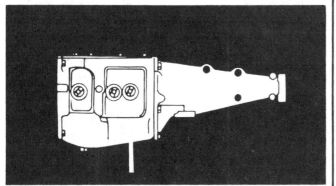

433 Ford T & C
10 bolt top cover

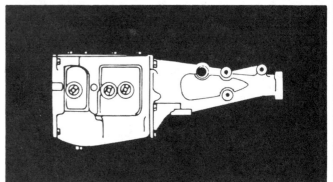

434 Ford T & C
10 bolt top cover

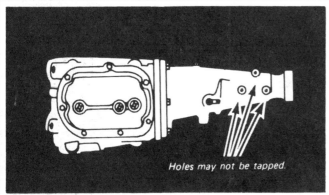

Holes may not be tapped.

441 Saginaw
7 bolt side cover

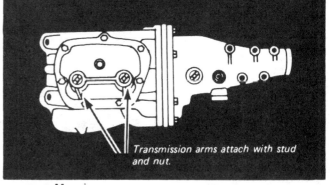

Transmission arms attach with stud and nut.

451 Muncie
7 bolt side cover used from 1963-68

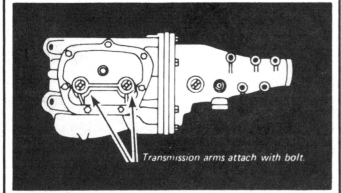

Transmission arms attach with bolt.

452 Muncie
7 bolt side cover 1969 and later

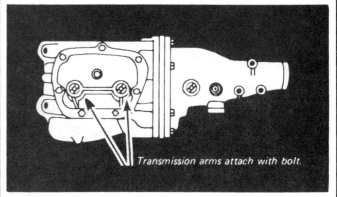

Transmission arms attach with bolt.

453 Muncie
7 bolt side cover 1970, 454 Chevelle

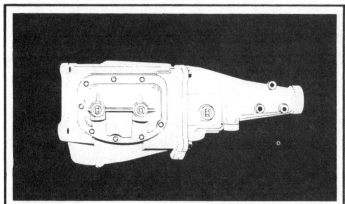

454 Borg-Warner T-10, (AS-9) 1974 and later
9 bolt curved bottom side cover

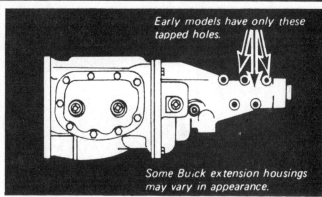

Early models have only these tapped holes.

Some Buick extension housings may vary in appearance.

410 Borg-Warner T-10, Super T-10 (AS-3)
9 bolt curved bottom side cover

3 speed transmissions

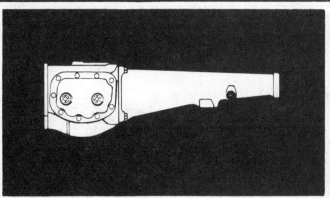

Holes may not be tapped.

311 Borg-Warner T-16
9 bolt side cover; synchronized first gear

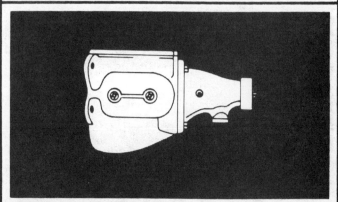

312 Borg-Warner T-85
9 bolt curved bottom side cover

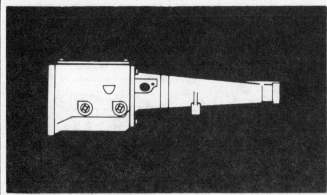

314 Borg-Warner T-86
6 bolt top cover

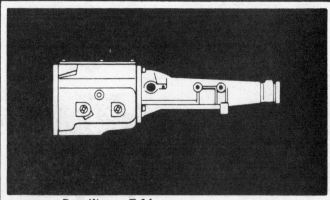

315 Borg-Warner T-96
4 bolt top cover

316 Borg-Warner T-14
Top cover

Some extension housings have parking brake assembly.

321 Chrysler Product
6 bolt top cover

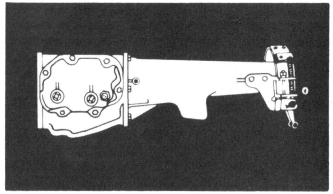

322 Chrysler Product
6 bolt side cover

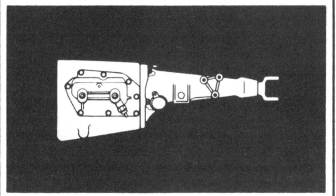

323 New Process
8 bolt side cover

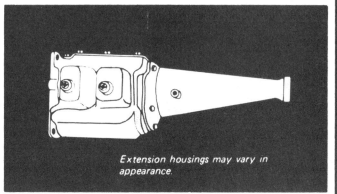

Extension housings may vary in appearance.

333 Ford
9 bolt top cover; first gear synchronized

334 Ford
9 bolt top cover; first gear synchronized

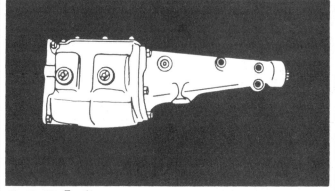

335 Ford
9 bolt top cover; first gear synchronized

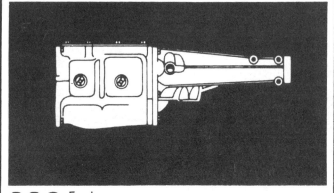

336 Ford
9 bolt top cover; first gear synchronized

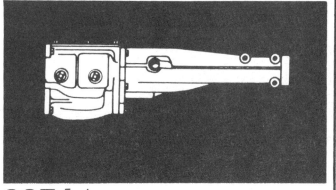

337 Ford
4 or 6 bolt top cover; first gear not synchronized

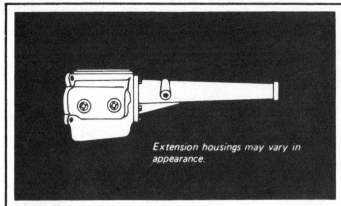

Extension housings may vary in appearance.

338 Ford
6 bolt top cover; first gear not synchronized

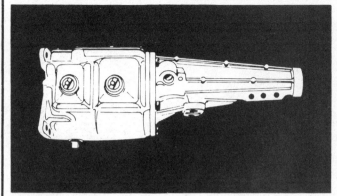

339 Ford
9 bolt top cover; first gear synchronized

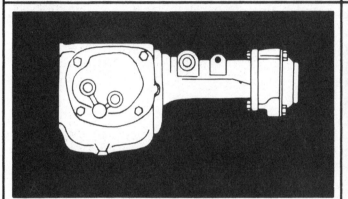

340 Chevrolet
4 bolt side cover with torque-tube drive

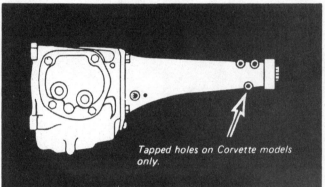

Tapped holes on Corvette models only.

341 Chevrolet
4 bolt side cover with round gear selector shafts

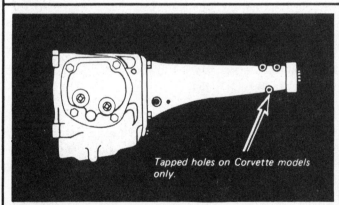

Tapped holes on Corvette models only.

342 Chevrolet
4 bolt side cover with keyed selector shaft

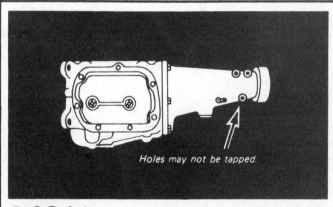

Holes may not be tapped.

343 Saginaw
7 bolt side cover; first gear synchronized
CAUTION: This transmission closely resembles 351 Muncie.

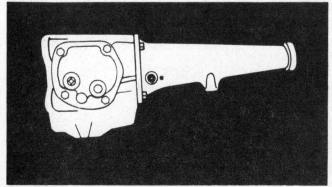

344 Chevrolet   4 bolt side cover; one round
selector shaft; one keyed selector shaft

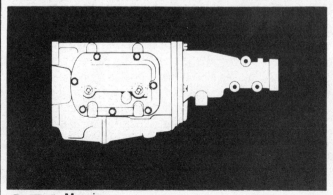

351 Muncie
7 bolt side cover; first gear synchronized
CAUTION: This transmission closely resembles 343 Saginaw.

When you purchase an engine or transmission like this you don't know the condition of the vehicle it came out of, inspect all parts very carefully for hidden damage.

used with Hemi's and 440's were built for extra strength and are more desirable for racing. They are identified by an 18-spline front input shaft. All others have 23-spline input shafts. Both types of transmissions have both aluminum and cast iron cases, and the cases are all interchangeable.

CHEVROLET REAR ENDS

There are two types of General Motors Salisbury rear ends, a 10-bolt ring gear and a 12-bolt ring gear. The 12-bolt variety is much preferred for high performance. This is easy to spot because the 12-bolt housing inspection cover has twelve cap screws. The 10-bolt rear end has ten cap screws in the inspection cover.

FORD REAR ENDS

The Ford 9-inch ring gear differential is the standard rear end for most racing applications. This rear end is widely available, easy to work on, very strong, and there are a number of ratios available for it. The easiest way to spot the 9-inch rear end first of all is its removable carrier. However, there are four different removable carrier Ford rear ends — 8-inch, 8 3/4-inch, 9-inch and 9 3/8-inch ring gears. Of these four, two can be immediately identified as NOT being the 9-inch. The 8-inch ring gear carrier is obviously small, and it is found in lighter-weight light duty vehicles. The 9 3/8-inch carrier has a pinion snubber cast on the housing above the companion flange as an integral part of the carrier, where as the 9-inch housing uses a bolt-on snubber. When the choice is narrowed down to a suspected 9-inch ring gear rear end, the carrier can be pulled and the ring gear measured.

There is a stamped metal product identification tag attached by a carrier housing bolt on all 1963 and later Ford axle housings. The code stamped on that tag will help you determine the ring gear diameter, application and ratio. A Ford parts man can make the interpretation for you, or the wrecking yard should have a Hollander Interchange Manual which will identify all the codes.

Once you have everything narrowed down to the 9-inch ring gear carrier, there are still some differences to sort out. First of all, there are two different size pinion carrier bearings — 2-57/64-inches O.D. and 3-1/16-inches O.D. The larger bearing is preferred for racing use. Secondly, there have been four different sizes of pinion gear rear roller bearings. All work equally well for any application, but if you have a number of rear ends you should try to have all of them accept the same size bearing so you need not stock several sizes of bearings for repair.

There are several different carrier housings which have been used with the 9-inch rear end. The most desirable is the high nodular cast iron case which is easily identified with an "N" cast on it. There are also high nodular iron cases which are acceptable which have the cross hatch ribs cast in the housing, but not the characteristic "N."

INSPECTING PARTS AT THE WRECKING YARD

Conduct as thorough an inspection as possible BEFORE buying your parts. It will save both you and the yard a lot of trouble and hassle.

Be sure to find out in advance what the yard will guarantee, especially if major components are mechanically sound and free of cracks. Also be sure to get a written receipt for your parts, and if possible, get the guarantee in writing on the receipt. Then as soon as you get your parts home, tear them down and thoroughly inspect them for damage.

INSPECTING THE ENGINE

Check the engine first for obvious problems such as

cracks, broken bosses, stripped threads, etc. If the engine is stil in a wrecked car, you can, through a little detective work, figure out how much use, wear and care the engine has received. Does the car have high or low mileage? Does the car appear to have had good care before the wreck? If so, chances are the engine has had also. Check the oil in the pan and the filter. Look for water in the oil and metal chips in the oil. Pull the pan and look at the bearings. Check the block and heads for obvious cracks and obvious oil or water leakage problems.

If you are buying an engine out of a wrecked car, look for these danger signals: an engine with a damaged crank dampner caused by a front end blow could mean the main caps or main webbing in the block have been damaged. An engine with a damaged transmission could mean the same thing. An engine with a ripped or missing oil pan or oil filter could mean serious damage to the block or other internal parts. And, do not buy an engine out of a car which has had a fire — the intense heat of a fire will weaken and warp cast iron parts.

INSPECTING THE TRANSMISSION

This subject is covered much more thoroughly in the transmission chapter elsewhere in this book. But, briefly, look for obvious cracks and problems with the case. Many times a problem crack in a transmission case is not obvious to the eye, but it can be suspected when there is an accumulation of moisture or oil and dirt in a certain area.

Put the transmission in each gear and rotate the input shaft. Everything should turn freely through each gear. Then remove the cover and inspect the gears, synchronizers and shifter forks.

INSPECTING THE REAR END

Start with an inspection of the case, looking for cracks. Then pull the carrier out and check the teeth — are they scored, broken or chipped? Also, smell the rear end grease, checking for a burnt smell. If it does smell burnt, pass on the part, because it indicates a bad pinion bearing which you cannot see. Don't worry about recognizing the burnt grease smell — if it is, you'll definitely know it.

To determine the rear end ratio, count the number of teeth on the ring gear and the pinion gear. Divide the number of ring gear teeth by the number of pinion teeth.

Stay away from engine and transmission that came out of a vehicle which has suffered a severe front end impact. Internal parts and block could very well be cracked.

Hardware

FASTENER HARDWARE

Fasteners are not glamorous parts. They don't impress many people because they don't make the car go faster. But they are relatively small parts which are being called upon to hold together big parts and withstand severe stresses, so you should know how to select and treat them right in order to have the maximum insurance that those big parts don't fall apart. Keep in mind that threaded fastener failure is a frequent cause of race cars not finishing a race — or even worse consequences. Treat them right and they'll treat you right.

GRADING

There are two popular grading systems for threaded fastening hardware: the SAE grade system, and the AN, MS and NAS military standards system.

The SAE grading system uses a series of ten numerals to indicate the various strengths of bolts. The various grades are marked differently on the head of the bolts to readily identify the strength of the bolt. Refer to the accompanying chart for identification of the markings.

The military standards for hardware far exceed the SAE grading system, most notably because hardware produced under these specifications must meet rigorous military specifications and inspection. The AN stands for Air Force/Navy standards, the MS for Military Specification and the NAS for National Aerospace Standard. Regardless of the letter designations, the standards are uniform.

The military specification hardware is more highly regarded than the commercial SAE graded hardware because the AN and MS pieces are manufactured to much closer tolerances, their surfaces are truer, their strength and hardness are more consistent and they are much more closely inspected. Sounds like the AN and MS hardware should cost more, right? No, thanks to our military's purchasing policies, surplus AN and MS hardware is readily available throughout the country at reasonable prices, usually much less than the cost of the SAE Grade 5 and Grade 8 hardware. To find surplus AN and MS fasteners in your area, look in the Yellow Pages under "Aircraft Parts and Supplies," "Industrial Fasteners," and "Government Surplus."

When you find a source of parts, just go in knowing what you are looking for. Generally for heavy duty race car service, you will be using the steel AN3 through AN20 series of aircraft machine bolts. The series numbers designate the grip and length of the bolt. The minimum tensile strength of AN3-20 series is 125,000 psi with a minimum shear strength of 75,000 psi.

If you are using the SAE graded hardware, use only Grade 5 and Grade 8 hardware for critical applications in your race car. Use Grade 8 where there are tension loads only on the bolts. Use Grade 5 where there is a ductility requirement needed to absorb shocks (for example, suspension components). Grade 8 is highly heat treated and is therefore more brittle. It will give less service where continual shock input has to be absorbed.

TORQUE

Knowing how much torque to apply to a fastener is just as important as knowing how to select the correct fastener. Torque specs for every type of bolt are usually expressed in foot pounds (ft/lb) and indicate the "proof load" of the bolt. Proof load is the amount of torque which can be safely applied to a bolt and nut without stretching it. Lubrication on a bolt has more to do with torque accuracy than you might think. If a bolt is coated with an anti-seize compound, it will require 45% less torque than a completely dry bolt. Most torque charts show torque specs for oiled bolts.

Grade Marking	Specification	Material	Bolt Size	Proof Load	Tensile Strength
	SAE—Grade 0	Steel	1/4 thru 1-1/2	–	–
	SAE—Grade 1	Low Carbon Steel	1/4 thru 1-1/2	33,000	60,000
	SAE—Grade 2	Low Carbon Steel	1/4 thru 3/4 Over 3/4 thru 1-1/2	55,000 33,000	74,000 60,000
	SAE—Grade 3	Medium Carbon Steel, Cold Worked	1/4 thru 1/2 Over 1/2 thru 5/8	85,000 80,000	110,000 100,000
	SAE—Grade 5	Medium Carbon Steel, Quenched and Tempered	1/4 thru 3/4 Over 3/4 thru 1 Over 1 thru 1-1/2	85,000 78,000 74,000	120,000 115,000 105,000
	SAE—Grade 5.1	Low or Medium Carbon Steel, Quenched and Tempered with Assembled Lock Washer	Up to 3/8 incl.	85,000	120,000
	Formerly SAE—Grade 6	Medium Carbon Steel Quenched and Tempered	1/4 thru 5/8 Over 5/8 thru 3/4	110,000 105,000	140,000 133,000
	SAE—Grade 7	Medium Carbon Alloy Steel, Quenched and Tempered, Roll Threaded After heat treatment	1/4 thru 1-1/2	105,000	133,000
	SAE—Grade 8	Medium Carbon Alloy Steel, Quenched and Tempered	1/4 thru 1-1/2	120,000	150,000
	Exceeds SAE—Grade 8	Chrome-Nickel-Molybdenum Electric Furnace Alloy Steel Quenched and Tempered	1/4 thru 1-1/2	165,000	185,000

SAE GRADE MARKINGS FOR STEEL BOLTS

Society of Automotive Engineers (SAE) bolt head markings indicate relative strength: the more lines there are, the tougher the bolt is. "Proof load" is the amount of load that can be safely applied without stretching the bolt. "Tensile strength" means the amount of load required to cause bolt failure.

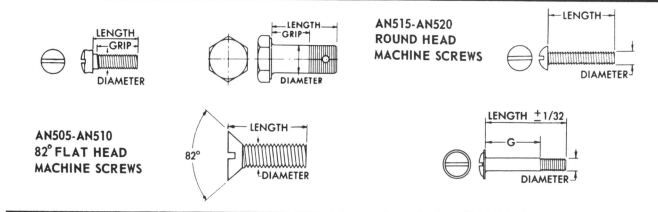

BOLT DIAMETER

Both the SAE Graded and AN hardware are available in a wide range of common fractional-inch diameter sizes. The use of these is fairly obvious, with the choice of a diameter size usually dependent upon the size of a bolt which it is replacing or the size of a hole through which the bolt must pass.

BOLT LENGTH

The bolt length is that length measured from under the head to the end of the shank. The grip is the portion of the shank length from under the bolt head to where the threads begin, which is actually the thickness of material the bolt is designed to hold or grip. The grip is the most important factor in selecting bolt length.

The grip should pass through a hole with the exception of the first two threads on the shank which should remain in the hole. Because the last two threads of a bolt (as well as the first two threads) are considered to be imperfect, they should not be depended upon for proper torque to contain the structure. The last two threads should not enter the nut. A washer can be used to space between the structure and the nut.

Why not use a fully threaded shank bolt to pass through a hole in a structure? First of all, the load bearing on the threads will crush the threads. Secondly, the threads can act as a milling machine on the structure under movement or vibration, and the bolt will distort the hole.

Bolts which enter tapped holes should be selected so that the bolt will not bottom out in the hole, and so that the last two threads of the bolt will not enter the hole. Use a washer to take up the spacing of the last two threads.

Whenever you run into a problem of grip length versus overall length, look to the AN and MS hardware for a solution to your problems. The AN hardware has a far greater variety of grip lengths available than the commercial SAE graded fasteners.

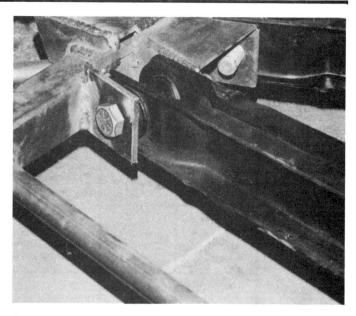

Suspension locating hardware like this should be Grade 5. Grade 8 is too brittle. Better than Grade 5 is AN hardware.

THREAD CARE

Bolt threads should be carefully inspected for damage or burrs. A die run down the threads can usually clean up the problems. Never clamp a bolt in a vise by the threads. Never grind on a bolt — the frictional heat may change the heat treat and, thus strength qualities of it. Never clean bolt threads with a wire brush or wire wheel — damage or thread deterioration will result. Use a chemical cleaner such as Loctite's Klean N' Prime or solvent.

THREADS

In all types of bolts, there are two types of threads — coarse and fine. Coarse-threaded bolts should only be used in tapped holes of material softer than the bolt (such as in castings) or where the application calls for frequent assembly and disassembly (such as wheel studs) where cross threading could become a problem. All other situations should require the useage of fine threaded bolts because their strength is at least ten percent greater given the same size and tensile strength bolt.

BOLT FINISH

Carbon steel and alloy steel bolts generally come with two types of exterior finishes — cadmium plating or black oxidizing. The cad plated bolts are fine for exterior application where a little more corrosion resistance as well as better appearance is desired, but don't use them in internal applications. How would you like to have cadmium flecks floating around in your engine oil or transmission lube? The problem is that cadmium plating is not permanent — nor perfect.

NUTS

No, not the type you race against, but rather the type which holds a bolt in place.

There are three types of nuts which you have to choose from for all bolt applications — plain, self-locking and castle or castellated. The self-locking type has some kind of locking device built in that keeps it and the bolt in place. The castle nut uses either a cotter pin or safety wire to keep everything in place. The plain nut just depends on blind faith and good luck to prevent vibration from loosening it.

The nylon-lined lock nut is much preferred over the all metal lock nut. The all metal lock nut holds its bond by distorting the bolt threads. Needless to say, this is intended as a permanent type of installation, and should never be considered for any use where disassembly and reassembly is a factor. Should it ever be disassembled, both the bolt and lock nut should be discarded and never reused. The nylon-lined lock nut should be considered for most all uses in a race car, except those where the bolt or nut is subjected to temperatures of 250 degrees F or more. At this temperature the structural integrity of the nylon is lost, and it will deform or melt.

There is one metal self-locking nut of the slightly

GLOSSARY OF FASTENER TERMS

AN — Air Force and Navy aircraft standard specification hardware.

Bolt — An externally threaded fastener which is tightened by torquing a nut on it.

Bolt size — is expressed as the diameter of the bolt shank. Bolts begin at ¼-inch diameter, and continue upward in 1/16-inch diameter increments through 5/8-inch diameter. From there on, the sizes are in 1/8-inch increments through 2 inches. Below ¼-inch diameter, the fastener is called a screw, and is designated by numeral sizes such as #2, #3, etc.

Edge distance — is the distance from the center of the hole to the nearest outer edge of a component.

Grade — A term assigned by the Society of Automotive Engineers (SAE) to the relative strengths of bolts.

Head marking — is a symbol found on a bolt head to indicate its strength.

Lock nut — is a nut which contains some type of device in itself to lock the nut in place on the bolt shank. The most common types are nylon insert and upset thread lock nuts.

MS — Military Standard specifications for hardware.

NAS — National Aerospace Standard for hardware.

Preload — is the force produced in a bolt as a result of torquing the bolt or a nut on it. Its opposite and equal reaction is clamping force. Preload stretches a bolt.

Shear strength — is the ability of a bolt to resist forces that try to cut it in two (forces perpendicular to the bolt shank).

Screw — is an externally threaded fastener similar to a bolt but under ¼-inch shank diameter.

Tensile strength — is a bolt's ability to resist forces pulling it apart along its length. Tensile strength can be expressed as both yield strength and ultimate strength.

Threads — come in two types of bolts: coarse and fine. Coarse threads are used when thread strength is more important than shank strength. Fine threads are used when shank strength in tension (tensile strength) is more important than thread strength.

Ultimate strength — is the form of tensile strength which is required to actually break a bolt.

Yield strength — is the point at which forces acting in tension (pulling apart) begin to stretch a bolt without increasing the load on it. When a bolt has yielded, it has been over-torqued and will not return to its original length.

deformed type which does **not** do damage to bolt threads, and is reuseable up to 6 times. It is the Bowman Products Bowmaloy nut. It is highly recommended for race car use.

The general range of readily available nuts are not graded for strength as the bolts are. Rather, nut strength is increased by the thickness of the nut, which in turn means that more nut strength is accomplished by the addition of more threads. When you are selecting a nut for a bolt which has critical forces in tension, be sure to select a thick nut to finish the job properly.

When using any type of nut — plain or locking — be sure that at least two threads of the bolt are protruding through the nut after it has been torqued into position. Remember, the last two threads on the bolt are considered to be imperfect.

WASHERS

The major reason for using a washer under both the bolt head and nut is to prevent stress concentrations on the base structure and to prevent galling the structure. The washer under the bolt head also prevents stress concentrations in the bolt fillet radius under the head.

Lock washers used with plain nuts should not be considered as a suitable replacement for a regular lock nut. The lock washer cannot be counted on in critical situations.

BOLT TORQUING

The torquing of a bolt is done to build a proper preload or clamping force. This prevents bolt loosening and prevents thread deformation. It is far easier to tell you that **all** bolts should be torqued than to have to go through the entire car and actually do it. But you should have a bolt torque chart, be familiar with it, and at least torque all critical installations. You'll never be sorry.

Torque charts may indicate the torque value for dry bolt threads, oiled bolt threads, or both. If the chart does not specify which situation, assume the torque ratings are for dry threads. The most common type of thread oil is Anti-Seize, which prevents thread galling, corrosion and seizing. If Anti-Seize is not used, select a light machine oil. Torque values for these two thread lubricants are 20 percent less than those for dry threads because the oil lessens metal-to-metal thread friction. If heavier oils are used, torque values are reduced even more. For example, if SAE 40 grade motor oil is used as a thread lube, the torque value should be reduced by 35 percent from the dry thread specification. The very slippery moly-based greases, when used as a thread lube, would require the torque value to be reduced by 50 percent.

SAFETY WIRE

Safety wiring is the best way to be sure your bolt is going to stay put. It is much better than a lock washer or even self-locking nuts. Safety wiring takes a special tool (safety wire pliers) and it takes only a little practice to get the feel of it. Be sure to wire a bolt so that it will tighten rather than loosen the bolt as vibration works it. Use .032-inch diameter stainless steel wire for safety wiring. Stainless steel is softer, and thus more malleable, than carbon or alloy steels. We have listed here for you the most common items which should be safety wired on any race car:

1. Rear end drive plate cover bolts
2. Fan attachment bolts
3. Transmission drain plug
4. Rear end drain plugs and vent plugs
5. Pinion retaining bolts on quick change
6. Fuel cell drain plug
7. Steering arm retaining bolts
8. Retaining bolts for disc brake brackets
9. Retaining bolts for disc brake calipers
10. Retaining bolts for rotors and hats
11. Retaining bolts for drum brake backing plates
12. Starter motor bolts
13. Header bolts
14. Safety wire, when doubled over, can replace a lost cotter key **in an emergency**

TORQUE CHART

Bolt Diameter [Dry Threads]

SAE Grade	1/4	5/16	3/8	7/16	1/2	9/16	5/8	3/4	7/8	1	1-1/8	1-1/4	1-3/8	1-1/2	1-5/8	1-3/4	1-7/8
0,1,2	6	12	20	32	47	69	96	155	206	310	480	675	900	1100	1470	1900	2360
3	9	17	30	47	69	103	145	234	372	551	872	1211	1624	1943	2660	3463	4695
5	10	19	33	54	78	114	154	257	382	587	794	1105	1500	1775	2425	3150	4200
6	12	24	43	69	106	150	209	350	550	825	1304	1815	2434	2913	3985	5189	6980
7	13	25	44	71	110	154	215	360	570	840	1325	1825	2500	3000	4000	5300	7000
8	14	29	47	78	119	169	230	380	600	700	1430	1975	2650	3200	4400	5660	7600

All torque is expressed in foot-pounds.

DRILL, TAP SIZE CHART

DECIMAL EQUIVALENTS

DRILL SIZE	DECIMAL	DRILL SIZE	DECIMAL	DRILL SIZE	DECIMAL	DRILL SIZE	DECIMAL
1/64	.0156	7/64	.1094	15/64	.2344	7/16	.4375
80	.0135	35	.1100	B	.2380	29/64	.4531
79	.0145	34	.1110	C	.2420	15/32	.4688
78	.0160	33	.1130	D	.2460	31/64	.4844
77	.0180	32	.1160			1/2	.5000
76	.0200	31	.1200	1/4	.2500		
75	.0210			F	.2570	33/64	.5156
74	.0225	1/8	.1250	G	.2610	17/32	.5313
73	.0240					35/64	.5469
72	.0250	30	.1285	17/64	.2656		
71	.0260	29	.1360	H	.2660	9/16	.5625
70	.0280	28	.1405	I	.2720		
69	.0292			J	.2770	37/64	.5781
68	.0310	9/64	.1406	K	.2810	19/32	.5938
1/32	.0312	27	.1440			39/64	.6094
		26	.1470	9/32	.2812		
67	.0320	25	.1495			5/8	.6250
66	.0330	24	.1520	L	.2900		
65	.0350	23	.1540	M	.2950	41/64	.6406
64	.0360					21/32	.6562
63	.0370	5/32	.1562	19/64	.2969	43/64	.6719
62	.0380	22	.1570				
61	.0390	21	.1590	N	.3020	11/16	.6875
60	.0400	20	.1610				
59	.0410	19	.1660	5/16	.3125	45/64	.7031
58	.0420	18	.1695			23/32	.7188
57	.0430			O	.3160	47/64	.7344
56	.0465	11/64	.1719	P	.3230		
		17	.1730	21/64	.3281	3/4	.7500
3/64	.0469	16	.1770				
55	.0520	15	.1800	Q	.3320	49/64	.7656
54	.0550	14	.1820	R	.3390	25/32	.7812
53	.0595	13	.1850			51/64	.7969
				11/32	.3438		
1/16	.0625	3/16	.1875			13/16	.8125
				S	.3480		
52	.0635	12	.1890	T	.3580	53/64	.8281
51	.0670	11	.1910			27/32	.8438
50	.0700	10	.1935	23/64	.3594	55/64	.8594
49	.0730	9	.1960				
48	.0760	8	.1990	U	.3680	7/8	.8750
		7	.2010				
5/64	.0781			3/8	.3750	57/64	.8906
		13/64	.2031			29/32	.9062
47	.0785			V	.3770	59/64	.9219
46	.0810	6	.2040	W	.3860		
45	.0820	5	.2055			15/16	.9375
44	.0860	4	.2090	25/64	.3906		
43	.0890	3	.2130			61/64	.9531
42	.0935			X	.3970	31/32	.9688
		7/32	.2187	Y	.4040	64/64	.9844
3/32	.0937						
		2	.2210	13/32	.4062	1	1.000
41	.0960	1	.2280				
40	.0980	A	.2340	Z	.4130		
39	.0995			27/64	.4219		
38	.1015						
37	.1040						
36	.1065						

TAP DRILL SIZES

THREAD	DRILL	THREAD	DRILL
0-80	3/64	1/2-20	29/64
1-64	NO. 53	9/16-12	31/64
1-72	NO. 53	9/16-18	33/64
2-56	NO. 50	5/8-11	17/32
2-64	NO. 51	5/8-18	37/64
3-48	NO. 47	3/4-10	21/32
3-56	NO. 45	3/4-16	11/16
4-40	NO. 43	7/8-9	49/64
4-48	NO. 42	7/8-14	13/16
5-40	NO. 38	1-8	7/8
5-44	NO. 37	1-12	59/64
6-32	NO. 36	1-14	15/16
6-40	NO. 33		
8-32	NO. 29		
8-36	NO. 29		
10-24	NO. 25		
10-32	NO. 21		
12-24	NO. 16		
12-28	NO. 14		
1/4-20	NO. 7		
1/4-28	NO. 3		
5/16-18	F		
5/16-24	I		
3/8-16	5/16		
3/8-24	Q		
7/16-14	U		
7/16-20	25/64		
1/2-13	27/64		

TAPER PIPE

THREAD	DRILL
1/8-27	R
1/4-18	7/16
3/8-18	37/64
1/2-14	45/64
3/4-14	59/64
1-11 1/2	1-5/32

STRAIGHT PIPE

THREAD	DRILL
1/8-27	11/32
1/4-18	7/16
3/8-18	37/64
1/2-14	23/32
3/4-14	59/64
1-11 1/2	1-5/32

SHEET METAL SCREW SIZES

DIAMETER	#4	#6	#7	#8	#10	#12	#14	5/16	3/8
DRILL SIZE	1/16	3/32	7/64	1/8	9/64	11/64	13/64	15/64	5/16

PIPE THREAD SIZES

1/8   1/4   3/8   1/2

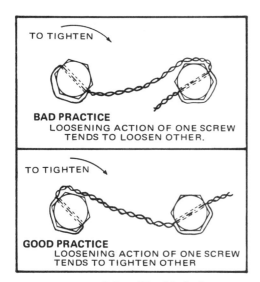

Safety Wire Methods

Drawing courtesy of Halibrand Engineering.

There are a variety of grades of rod end bearings. Be sure you know the quality and limitations of the part you buy, and if it will work for the application you have in mind.

CHEMICAL LOCKING AGENTS

The Loctite Corporation makes a range of chemical bonding agents which are designed to prevent bolts from loosening because of vibration. Their Stud N' Bearing Mount is the permanent type of Loctite, which requires two boys and a bull elephant to disassemble. Their Lock N' Seal is the removable type of thread sealant. Loctite's Wick N' Lock is a penetrating sealer which is applied on set screws once they have been adjusted to prevent them from backing off (for example, carburetor mixture adjustment screws). Loctite also makes another neat compound worth knowing about called PST, which stands for pipe sealant with teflon. It replaces teflon tape as a sealant in pipe threads — mighty neat stuff.

HELI-COILS

Heli-Coils are thread inserts which help you repair deteriorated threads caused by general wear, stripping, cross-threading or rust. These little goodies are definitely lifesavers for racers, and are generally available in kit form from parts stores.

The Heli-Coil threaded inserts can be used in situations where a threaded hole has been seriously damaged and the threads cannot be restored. Before the Heli-Coils are used, a serious attempt should be made to restore the threads with a tap first.

To use a Heli-Coil, drill the damaged hole oversize (follow the size directions in the kit). Then the special Heli-Coil tap is used to thread the hole. The hole is ready for the insert, so it is threaded into place using the special insertion tool provided. The tool grasps the insert by a tab on the bottom. Once the Heli-Coil is in place, the insertion tab is chipped and broken away with a screwdriver and hammer.

ROD END BEARINGS

Many times called Heim joints (which is a proprietary name brand), rod end bearings are one of the most effective methods of attaching arc-movement parts to a chassis.

Many people think that all rod end bearings are the same — regardless of the price paid for them. This could not be further from the truth however, and the ignorance of this fact can have severe consequences if poor quality bearings are installed in the wrong places on a race car.

GRADE

Rod end bearings are generally graded into two classifications — commercial and aircraft. The commercial grade is less expensive than the aircraft, but it is also less desirable because they are manufactured to wider tolerances and from inferior materials.

There are three parts to a rod end bearing — the body, the race, and the ball. All three must be constructed of compatibly strong materials. To cut costs, a car builder can find a rod end bearing with a good quality body material, but the manufacturer may have skimped on the ball or race material in order to cut costs. You must be aware of this

Helicoil thread inserts are a lifesaver for many cast parts where threads have been stripped or worn.

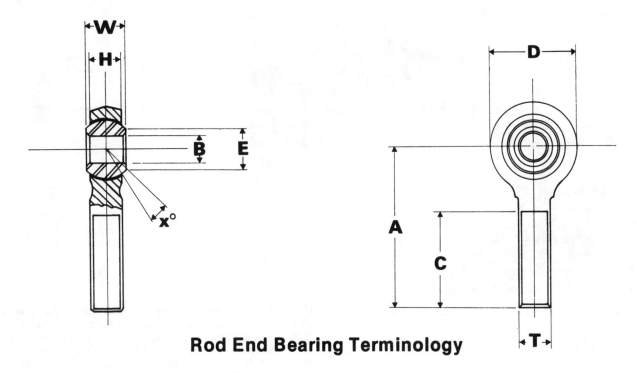

Rod End Bearing Terminology

CODE	DESCRIPTION
A	Body Length
B	Bore Diameter
C	Shank Length
D	Body Width
E	Lip Diameter
H	Body Thickness
T	Thread Diameter
W	Ball Width
X	Misalignment

when purchasing bearings. In order to be a quality rod end bearing, the body material must be as strong as the ball material, and the liner must be strong enough to handle the impact and compression fed into it from the other pieces.

The cheapest and least desirable (in terms of strength) commercial rod end bearing ball is the oil impregnated ball. This may sound fabulous until you realize that the ball is only compressed metal shavings with a lubricant. The next step up in the commercial grade ball is the low carbon steel ball. This type of rod end bearing should not be used in a high stress application because a low carbon steel ball is generally coupled with a low carbon steel body and no race material is contained at all. The finest quality ball (which is naturally the aircraft grade), is made of heat treated stainless steel. Even here there are two grades of the stainless steel, 303 stainless and 440C stainless. The 440C stainless material has nearly double the strength of the 303.

The insert or race is important in a rod end bearing where the application places loads on it in two directions. This would be applications such as anti-roll bar arm mounts, upper and lower trailing arm links, and torque reaction rods. The stress on a Panhard bar or Watts linkage is always applied gradually — not in a sudden shock load — and it is always in one direction on the bearing.

The race is where the great wear in the rod end bearing occurs with a two directional loading. With wear, they eventually beat out and give a very sloppy bearing fit. Obviously, then, the aircraft quality bearings are the answer here. The aircraft bearing usually has a self-lubricating teflon insert (Heim calls theirs Uniflon). This greatly enhances the wear characteristics, and in addition has a characteristic of rejecting small grit and dirt to aid the race's longevity.

In applications on a race car where stress levels are not extremely high or critical — such as cable ends and throttle

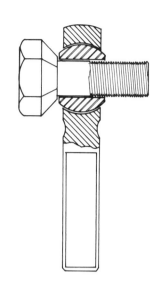

Upper left, rod end bearings should always have washers placed on either side of the ball. Above, the head of a bolt may be machined to add to the amount of misalignment available with a rod end bearing, yet attaching bolt clearance is still maintained. Below, two female shank bearings like this can be used with a threaded shank as an anti-roll bar arm attachment link.

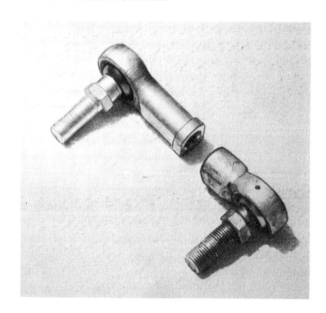

or clutch linkage — the commercial rod end bearings will fill the need just fine. Bearing races in these applications could be aluminum bronze, cadmium plated low carbon steel or fiberglass reinforced nylon with molybdenum disulfide added. The aluminum bronze and fiberglass reinforced nylon would be self lubricating, which is good for an application like a throttle linkage end or clutch linkage (because you are going to neglect the lubrication at these points anyway).

There are three materials commonly used for the body of the rod end bearing — low carbon steel (for commercial bearings), heat-treated stainless steel and chrome moly (4130) steel. The higher tensile strength materials are important in the body to prevent material stretch, which deforms the race and leads to a sloppy bearing fit as well as hastening wear.

Misalignment is also a consideration in the choice of rod end bearings. Misalignment is the maximum angle that the ball will travel before the head of the bolt contacts the bearing body. Because of numerous design factors, rod end bearings are available with a misalignment factor ranging from a low of six degrees all the way up to seventeen degrees.

If you have a low angle misalignment bearing and wish to increase the movement of the ball, use a 1/8-inch to 1/4-inch thick washer-type spacer on each side of the ball to add more room to its movement within the body without binding. The spacer diameter should be equal to the outside diameter of the lip of the ball, which will almost double the amount of misalignment.

When installing rod end bearings, be sure the mounting will not put any bending loads on the bearing. It does not take much at all in the way of bending loads to fatigue a rod end. Additionally, do not chrome plate rod end bearings, or any suspension piece which is stressed. The chroming

process will change the temper of the material, creating weak spots (called hydrogen embrittlement). The chrome plating can also hide a small crack. If you are looking for an attractive rod end bearing in a highly stressed useage, use the aircraft quality stainless steel part. If you need an attractive bearing in a low stress area, use a cadmium plated low carbon steel bearing.

On female shank rod end bearings, a bolt or threaded shank should extend as far into the bearing shank as possible to avoid excessive stress on the shank. No more than 3/8's of the total number of threads on a bolt should extend outside the bearing female shank. More exposed threads than this will cause the rod end shank to stretch and elongate the ball housing. This can cause wear and damage to the ball and race. To help prevent shank thread deformation and stretch on both male and female rod ends, use a jam nut against the shank.

For stock car racing application, we make these size recommendations for the use of rod end bearings (sizes refer to bore diameter): anti-roll bar, 1/2-inch; Panhard bar 3/4-inch; Watts linkage 5/8-inch; throttle linkage—at your discretion; A-arms 3/4-inch for cars weighing over 2,300 pounds. For cars under 2,300 pounds, 5/8-inch bearings can be used for A-arms. If you ever run into an application where you are undecided between two different sizes, always opt for the larger size. The difference in weight will be practically nil, the difference in cost is minimal, but the difference in strength is quite appreciable.

Two more uses for rod end bearings, above as a stabilizer for a steering shaft, and left as inner pivot attaching points for A-arms.

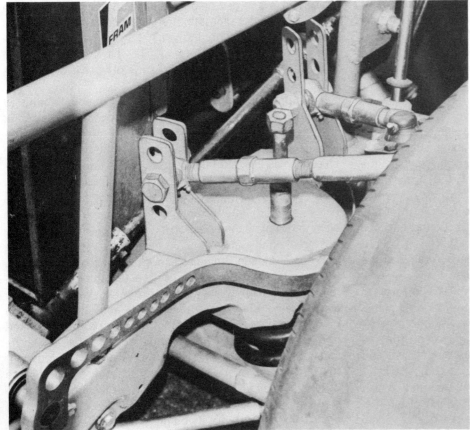

Spindles and Steering

We will start off by clarifying the terminology which we will use in this chapter. Actually, the spindle is two pieces — an upright or knuckle which is supported by the upper and lower ball joints, and the spindle itself or bearing carrier shaft, which is supported by the knuckle. In this book, we will call the complete two piece unit a spindle. If we are making reference to one of the two separate pieces, we will refer to the knuckle or the spindle shaft.

As we progress through our discussion on spindles, we will make specific comparisons of varying elements of some of the most common spindles being used in race cars today. These are the Holman and Moody racing spindle, the Speedway Engineering racing spindle, the Stock Car Products racing spindle, the GMC 3/4-ton pickup truck spindle part number 3935178, the 1958 to 1960 Mercury Parklane passenger car spindle (which is identical to the 1966 and 1967 Ford Galaxy with disc brakes or 1970 Ford Galaxy police and taxi spindle with drum brakes), and the 1961 to 1968 Cadillac passenger car spindle.

All of these spindles are produced as one-piece forgings in terms of the knuckle and spindle shaft — except the Cadillac spindle. All of the spindles also have spindle arms as an integral part of the spindle forging except for the Speedway Engineering, Stock Car Products, Cadillac, and GMC truck.

The Cadillac spindle has some unique features which make it attractive for a racing application. First of all, it is widely available in wrecking yards at a reasonable price.

Secondly, it is very strong in that it is a two-part forging. The knuckle is forged as one piece, with the grain going the right direction for ultimate strength. The spindle shaft is a separate forging, with the grain flowing in the opposite direction of the knuckle forging. The spindle shaft is driven in and shrunk fit, to make an extremely strong, yet lightweight unit.

The upper ball joint hole of the Cadillac spindle is fitted with an eccentric tapered bushing. The eccentric is used in the passenger car application to change camber and caster simultaneously. But in a race car, it can be used to change steering axis inclination. The least inclination available is 7 degrees, and the most is 10. The eccentric bushing, because it is tapered, is held very tightly in place by the nut on the ball joint stud, so welding is not required to hold the eccentric in place. Welding on it would only serve to ruin the heat treat on the knuckle.

When comparing spindles to decide which one to use in a race car, there are several critical elements to consider. The first is the steering arm length from the lower ball joint centerline to the tie rod end centerline. The longer this distance, the slower the steering ratio will be. As the tie rod end is placed closer to the spindle or lower ball joint, the steering ratio is sped up. If there is a spindle which you want to use and it has a long steering arm distance, you can make other changes, however, to improve steering ratio. The steering arm can be cut, shortened and rewelded (which we do not recommend, but if you do it, be absolutely certain to

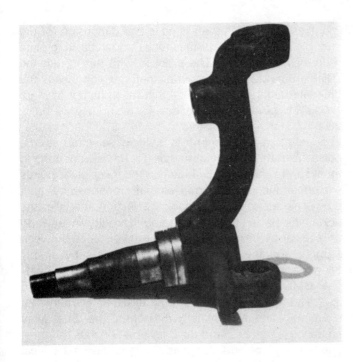

The Cadillac spindle referred to in the text. Above is the camber eccentric used to mount the upper ball joint in the Cadillac spindle.

have this done by a certified welder for safety's sake), a new arm can be fabricated (which would be quite costly) or a 16 to 1 ratio steering gear can be installed in the steering box. This would make steering sufficiently fast enough even with a long steering arm). Of the group of spindles we are comparing, the Mercury Parklane has a 7-inch arm, the Cadillac a 7.25-inch arm, the Speedway Engineering arm is adjustable for 5.25 or 6.625-inch length, and the Holman and Moody 6.25. The Cadillac, GMC, Speedway Engineering and Stock Car Products spindles all have bolt-on steering arms. This offers an advantage in that the arms can be reversed side-for-side to provide either front steering or rear steering on the spindles.

Another critical dimension of a spindle is its spindle shaft centerline to brake anchor pin hole centerline. We should add that this dimension is critical only on spindles which will be used in conjunction with the 11-inch diameter racing brake drums. In this case, the distance must be 4½ inches. All of the spindles in our comparison have this dimension except the Cadillac and the GMC. The GMC hole is only slightly higher, so it can be plugged with a piece of metal, the piece rosette welded, and a new hole redrilled in the knuckle to relocate the anchor pin. In the case of the Cadillac spindle, the hole is located too high and the flange surface is too shallow to allow the plugging and redrilling process. In this case, the Cadillac spindle can be used with

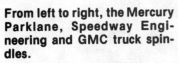

From left to right, the Mercury Parklane, Speedway Engineering and GMC truck spindles.

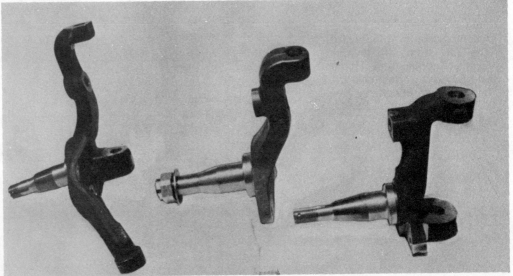

At top is the Speedway Engineering spindle steering arm. Below it is the Cadillac steering arm.

disc brakes (with which there is no critical dimension for the anchor pin hole heighth), or the stock Cadillac backing plate and drums can be used. The problem with this is that the Cadillac drums are 12 inches in diameter, which means the additional rotating weight will be that much harder to stop. The Cadillac spindle would be ideal for racing use with disc brakes.

The next critical element in spindle selection is the knuckle heighth. The heighth has to be tall enough, in comparison to the distance between the inner pivot points of the upper and lower A-arms, so that proper camber gain can be achieved. In terms of spindle selection, the taller the knuckle, the better. This allows you flexibility in building your race car by being able to adjust the distance between the upper and lower arm inner pivot points by simply moving the inner control shaft of the upper arm up or down.

The GMC spindle has the shortest knuckle at 7.875 inches. The Mercury Parklane is the tallest at 11.75 inches. The Holman and Moody is 8.75 inches, the Speedway Engineering is 9 inches, and the Cadillac is 9 inches.

The bearing size and side load rating is the next critical element. Naturally the Speedway Engineering, Stock Car Products and Holman and Moody racing spindles were especially designed for the most gruelling race track useage, so their bearing sizes are more than adequte. In fact, for the

A comparison of spindle sizes. Top row from left to right: GMC, Cadillac and Holman and Moody. Bottom row from left to right: Speedway Engineering, Mercury Parklane and Stock Car Products.

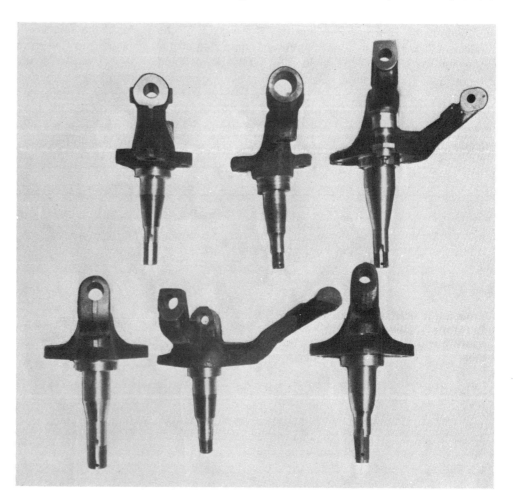

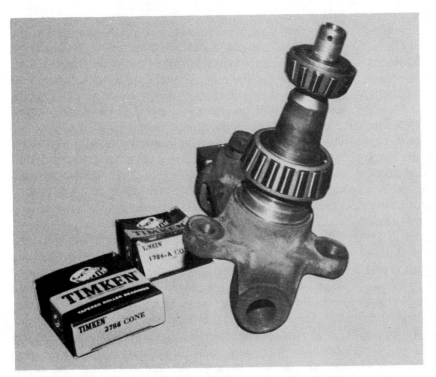

At left is the Cadillac spindle with Timken heavy duty bearings. The inner bearing is part number 2786, and the outer is part number 1784-A. Below is a comparison of the heavy duty bearings [top] against the standard bearings.

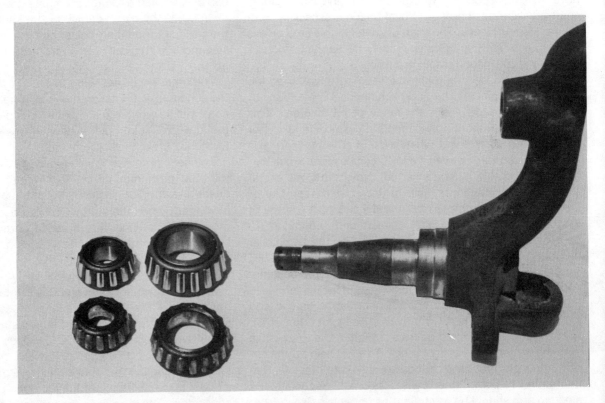

requirements of short track racing with a stock car, they are a classic case of overkill.

The basic requirements of the wheel bearings are to carry radial loads and thrust loads. The radial loads are the rotating forces and the downward weight imposed on the bearings. The thrust loads are the side loads generated by cornering. When looking at stock passenger car and light truck wheel bearings, the area they are most anemic in is the thrust load capacity. Thrust load carrying ability is gained by choosing roller bearings in which the rollers are both greater in diameter and longer in length.

Going hand-in-hand with bearing size is the proper maintenance schedule and preload of wheel bearings. Both a worn bearing and an improper preload adjustment can cause excessive rolling friction. This will affect the car adversely in two ways. First, the bearing operating temperature will be greater, and secondly, the overall speed of the vehicle will be reduced. Both increased friction and increased heat kill bearings. Proper maintenance means periodically checking the condition of the bearings, repacking them, and double checking the preload.

In the case of the GMC, Mercury Parklane and Cadillac spindles, larger capacity roller bearings are manufactured by Timken which have the same inside diameter as the stock application wheel bearings. The only requirement would be that a machined steel hub be used instead of a stock cast hub, which is much preferred any way because of safety.

The diameter of the spindle shaft at the outer wheel bearing is another critical element in spindle choice. As wheel offsets and tire tread widths grow, the forces imposed on the outer wheel bearing and spindle shaft in this area become much more critical. Too much wheel offset, combined with extreme cornering forces, have been known to break some spindle shafts in this area. Keep in mind, too, that proper bearing preload makes a big difference in preserving bearing and spindle shaft life.

Of all of the spindles we are comparing in this section, we have no knowledge of any problems existing with spindle shaft breakage at the outer wheel bearing, especially with the non-racing pieces. The shaft diameter in this area on the Cadillac and GMC spindles is virtually the same, in fact, as the Holman and Moody racing spindle.

Another selection factor is the steering axis inclination built into the spindle. The approximate ideal inclination for most stock cars today is in the range of 7½ degrees. This takes into consideration the effects of camber and caster settings, tire widths and wheel offsets. The steering axis inclination can vary betwen 7 and 10 degrees for a stock car, and be acceptable. Just a small adjustment in static caster and camber settings can help balance out any problems. Our comparison spindles all fall into the acceptable range of kingpin inclination, with the Holman and Moody at 10 degrees, the Speedway Engineering at 7½, the Stock Car products at 7½, the Mercury Parklane at 7½, and the GMC at 10½. The Cadillac spindle, with its eccentric top bushing, can be adjusted for either 7 or 10 degrees of inclination.

The next critical design consideration is the tie rod end placement in relationship to the lower ball joint. First of all, the tie rod end ball should be in as straight a line as possible with the center of the ball joint ball. This will help rid the car of bump steer problems (of course, there are other factors considered in bump steer causes as well). Secondly, the steering arm should be as straight as possible as seen from the top view to rid the suspension of Ackermann steering. Of the spindles we are comparing, the Holman and Moody and Mercury Parklane spindles are the worst offenders in the steering arm design category. The Speedway Engineering spindle was secifically designed to combat the steering arm problems we have outlined, as was the Stock Car

Products unit.

The final design consideration is the distance from the lower ball joint to the spindle shaft centerline. This distance directly affects the ground heighth clearance of the vehicle. The greater the distance between the lower ball joint and the spindle shaft, the lower the car will set, given a specific tire heighth. The greater distance is the most desirable. The Cadillac spindle has the greatest distance at 2.0625 inches. The Speedway Engineering and Stock Car Products spindles measure 1.6875 inches. The GMC is at 1.8125 inches, the Holman and Moody at 1.125 inches, and the Mercury Parklane measures .75 inches.

BUYING A USED SPINDLE

Naturally, the beauty of using a spindle such as the Cadillac, GMC or Mercury Parklane, is the relatively inexpensive cost of the unit. And, the cost gets to be very reasonable when you are able to purchase the parts from a wrecking yard. You can waste your money, however, if you do not check the parts thoroughly for structural soundness.

If you can, buy the spindles from the wrecking yard on the condition that they pass Magnaflux inspection. Then immediately have the parts Magnafluxed for cracks which can develop in the spindle shaft (especially toward the outer bearing), in the narrow areas of the knuckle just in front of the ball joint mounting holes, and the narrow areas of the steering arm (typically just in front of the tie rod mounting hole).

The other important thing to look at on a used spindle is the condition of the tapered ball joint holes. Many times the holes will be worn into a slightly oblong shape — especially the lower ball joint hole. If used in this way, the pounding forces will break the ball joint stud.

To check for a bad taper, thoroughly clean the bore that the stud taper mounts in. Then mount the ball joint and check the fit. The fit must be snug with no chance for movement when the stud is worked back and forth in the mounting hole. Only the threaded portion of the ball joint stud should protrude from the mounting hole.

HUBS

In building a race car, everybody is interested in saving as much money as possible. In a lot of places, stock passenger car or pickup truck parts can be substituted for the more expensive special racing parts, and do the job just fine. When it comes to hubs, however, this just is not so. Passenger car and truck hubs are made of cast iron, which is a material that will not get the job done properly in a racing application. The most serious problem is that the cast flange will break off the hub on the right front wheel. No matter how big and sturdy the cast iron hub is, this problem exists.

The only answer for a racing hub is a steel hub, machined from a one-piece steel billet. The initial investment in this piece will cost you a little more, but it is a piece which will be virtually unbreakable, and will last you for years on end. You

The stock cast hub, above, just will not get the job done safely on a race car. A steel billet hub, right, is the only way to go.

will be able to take it with you from car to car, as you build new racers. The most important aspect of the steel hub is safety. For just a few dollars more, you are buying a lot of insurance. Think of the extra cost you will incur should a cast hub at the right front break as you enter the first turn. What, then, would be the extra cost of new sheet metal, probably a new radiator, frame straightening, new A-arms and ball joints, plus a new hub and bearings and quite possibly the cost of medical services for the driver in case of injury? The risk is just not worth it. We hope you get the message we are trying to get across here — steel machined hubs do cost more money, but in the long run they save you money and possible injuries.

As you probably know and expect, the special racing application spindles from Holman and Moody, Speedway Engineering and Stock Car Products have machined steel billet hubs which come with them, either as a package or separately.

Speedway Engineering will machine a special steel billet hub for you for any application. They have on hand the patterns and blueprints required to readily make the Cadillac, GMC and Mercury Parklane hubs. If you have a special spindle for which you require a machined steel hub, supply them with the spindle and stock hub and they will make it for you.

WHEEL STUDS

The standard of the industry for wheel studs is a 5/8-inch diameter by 18 thread. The stud length for the front hubs is at least 2½ inches, with 2¾ and 3 inches sometimes used. The stud for the rear hubs is also a 5/8 by 18, at least 3 inches long, with 3¼ and 3½ -inch lengths also used.

The studs can be purchased from racing components suppliers, or you can find the ones which suit your purpose by going through a Dorman Products catalog. Dorman Products make a wide variety of replacement wheel studs as well as brake hardware. Check for their catalog at a parts store or brake hardware supplier.

When purchasing studs, pay attention to the length of the serrations on the grip under the head. They come in varying lengths. What you need are serrations which will not extend past the hub and drum when pressed in place, and preferably not past the hub.

The wheel studs take a lot of abuse on a race car. To make them last as they are intended, periodically clean the threads with a die, and spray a dry moly lubricant on the threads to prevent galling.

A variety of wheel stud lengths in the proper diameter and thread are available from Dorman Products.

BALL JOINTS

The accompanying tables indicate which ball joints to use with which spindles, for both upper and lower ball joint applications.

The ball joints were researched to find the best quality ball joints which would meet the following criteria: (1) As large a size as possible in both ball and stud, and yet still fit the existing tapers in the spindles without having to ream them. (2) Have a ball joint body which either bolts in place or screws in place, without the need of pressing the ball joint in place. The less press-in pieces with a race car, the better for ease of maintenance. (3) Wide availability and interchange.

The heaviest duty lower ball joint we have listed in the chart is the TRW number 10171N, which is a screw-in ball joint that has a Chrysler Imperial and Dodge truck application. This ball joint should be used on the bottom whenever possible as the loads imposed on the lower A-arms and lower ball joints are three times greater than

Spindle	Top Ball Joint	Bottom Ball Joint	
Holman & Moody	TRW 10262	10171N [1]	10209 [2]
1957 & later Ford Galaxy	TRW 10262	10171N	10209
GMC ¾-ton truck	TRW 10213	10171N [4]	10209 [4]
Speedway Engineering	TRW 10213	10171N	10209
Stock Car Products	TRW 10262	10171N	10209
Frankland	TRW 10213	10171N	10209
'61-'68 Cadillac	TRW 10195	10171N [3]	10209 [3]
'59-'60 Mercury Parklane	TRW 10262	10171N	10209

1. Screw-in body
2. Bolt-in body
3. Must use spacer washer
4. Must use adapter bushing

TRW Ball Joint	Source	Moog Interchange #
10213	1965-70 Chevy ¾-ton truck C-20, C-25 top	K-6024
10262	67-77 T-Bird, 1965-77 Ford Galaxy, Torino, and 1971-77 Pinto top	K-8212
10195	All 1965-70 Chevy top	K-5208
10171N	1970-73 Doge B-100/B-200 truck lower, 1964-66, 70-73 Imperial lower	K-727
10209	1972-1977 Ford Galaxy, bottom	K-8059

those on the top ball joints. This means that the bottom ball joint is the really critical component for taking the loads, and it should be heavy duty. It should also be checked periodically for wear. (Worn ball joints affect front end alignment and stability during cornering.) The screw-in ball joint (10171N) requires the use of a machined steel adapter, which is then welded to the A-arm. The adapter can be machined using the dimensions we have provided, or it is available for purchase from Speedway Engineering. The ball joint adapter can be used to an advantage to lower the chassis by welding it on top of the A-arm. This moves the lower ball joint further away from the spindle shaft centerline, which lowers the chassis. This is generally good for moving the chassis about an inch lower.

When the threaded ball joint adapter is fitted to the lower A-arm, screw the ball joint into it and position the assembly so that the ball joint is placed where it will not bottom out against its side during travel. Tack weld the adapter in the desired location, remove the balljoint, then finish the welding.

The Speedway Engineering threaded ball joint adaptor which welds onto the lower A-arm and accepts the screw-in type of ball joints.

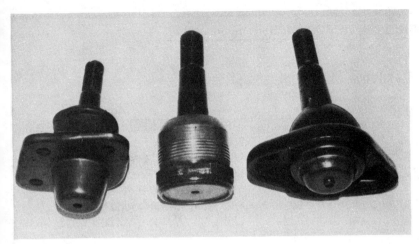

The three ball joints we discuss are from left to right the 10262, the 10171N, and the 10209.

A few notes are in order about using the 10171N ball joint with a couple of the spindles we have mentioned. The bottom ball joint hole and taper is too large on the GMC spindle to accommodate the ball joint stud, so a tapered adapter bushing must be used. Speedway Engineering makes this part, or you can have one machined by figuring out the spacing difference between the hole taper and stud taper which allows the ball joint stud threads only to protrude through the hole. This same adapter bushing would need to be used with the GMC spindle if the TRW 10209 ball joint were used.

When the 10171N or 10209 ball joints are used on the lower end of the Cadillac spindle, the spacer washer which is provided with the ball joint must be used because too much of the stud taper protrudes through the spindle hole. A flat spot on one side of the washer will have to be ground to allow the washer to seat flat on the knuckle (there is interference with the spindle shaft forging on the inside of the knuckle).

TIE RODS AND ENDS

You have two choices for tie rod ends — the type which will require reaming of the tie rod taper on steering components, or the type which will fit the stock components without modification. The choice boils down to a matter of strength, with the larger tie rod ends providing more strength than the stock sized ends.

Whether you need the added strength or not can depend on whether your car's spindles are front steering or rear steering. Front steering means the spindle steering arms point to the front of the vehicle, and the tie rods attach in front of the spindle. Rear steering would be just the opposite, or having the tie rods attach to the steering arms behind the spindles. More tie rod and tie rod end strength is required in cars which have rear steering spindles, because a bump or jolt to the right front wheel will most generally put the rear steering tie rods in compression, whereas it would put front steering tie rods in tension. Straight line road forces have the same affect on the tie rods. Tubular structures are much stronger in resisting tension loads than compressive loads.

With rear steering spindles, the ES258L and ES258R tie rod ends can be used. The "L" at the end of the part number means the shank is threaded for left-hand thread, and the "R" means right-handed thread. Another part number which interchanges with the ES258 is the ES261. One of each part number would be required to make up one tie rod. To construct the tie rod, assemble the chassis and have it setting at running heigth and close to the desired front end alignment. Install one tie rod end in the steering arm, and one in the steering center link. Measure from the tie rod end socket on one side to the socket on the other side. Subtract 1½ inches from that length, and you have the length required for the tie rod. Make the tie rod out of a

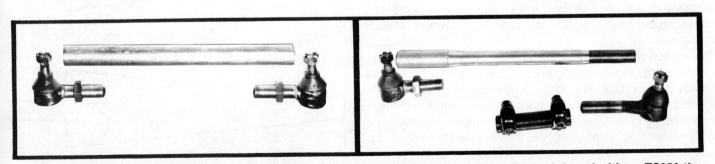

At left is a tubular tie rod arrangement with ES258 tie rod ends illustrated. At right is a solid steel tie rod with an ES258 tie rod end on the left and an ES150 tie rod end on the right.

SUGGESTED TIE ROD ENDS

Part #[1]	LH or RH Thread	Socket Dia. & Thread	Stud Taper High	Stud Taper Low	Stud Taper Length	Stud Thread	Application
ES150L	left	3/4 x 16	.701	.609	.736	9/16 x 18	71-76 Ford truck P350, P500
ES150R	right	3/4 x 16	.701	.609	.736	9/16 x 18	61-66 Ford truck F350, F500
ES258L[2]	left	3/4 x 16	.627	.543	.672	1/2 x 20	60-62 GMC 4WD K-1000
ES258R[2]	right	3/4 x 16	.627	.543	.672	1/2 x 20	1960 Ford truck F-100 & F250 4WD
ES261L[2]	left	3/4 x 16	.626	.543	.672	1/2 x 20	61-68 Dodge truck
ES261R[2]	right	3/4 x 16	.627	.543	.672	1/2 x 20	W100 & W250 4 WD

1 — TRW & Mood part members for the rod ends are identical
2 — Same taper as Monroe shock absorber tie rod end

length of 1-inch O.D., .156-inch I.D. 4130 steel tubing. Thread the ends for 3/4 by 16 threads, with left-hand thread at one end and right-hand thread at the other. Assemble the tie rod ends onto the rod with jam nuts.

The ES258 tie rod ends which we specified are large units, strong enough to take any racing punishment. The taper on the stud is the same taper as the Monroe tie rod end shock absorbers, which means that the ES258 is not going to fit right into Chevelle and other stock steering linkages without some work. (It will fit into larger Ford product stock steering linkages, such as 1971-77 Lincoln, 1971-77 big Ford models, and 1971-77 big Mercury models.)

To make the ES258 tie rod ends fit, the tie rod holes must be reamed. There is not a reamer made which has the exact same taper as the SAE specification of 1/8-inch of taper per inch. This taper specification means that if a hole starts with a 1-inch diameter, in one inch of hole length the diameter will shrink to 7/8-inch. The closest taper reamer made is called a 42R No. 2 carbon steel repairman's taper reamer (to find a dealer, look in the Yellow Pages under Machine Shop Supplies). This reamer has a taper of .135-inch per inch, versus the SAE specification of .125-inch per inch. The 42R No. 2 reamer can be used as it is, or if you are unhappy with the fit of the taper in the reamed hole, you can take the reaming tool to a toolmaker and have him reshape and resharpen it for you according to the drawing we have provided.

When you are using the taper reamer for tie rod holes, or shock absorber mounting holes, be very careful that you do not go too deep. Cut just a little material at a time then measure the tie rod end in the hole. Once you have driven the reamer too deep, you have ruined the part. Be sure to do the reaming in a drill press with the work part fixed solidly.

Another tie rod end which is widely used in racing applications is the ES150 L and R. This unit has the same 3/4-inch diameter threaded shank, but it has a larger base diameter tapered stud. Its extra size is **not** required, as many people think. The ES150 just requires more reaming of the

steering arm and center link taper holes.

If you are using a front steering car such as a Chevelle, etc., you can get by with using the stock tie rod ends and adjuster sleeve, which means no reaming will be required. Different years of Chevrolet, Camaro and Nova require various part numbers of inner and outer tie rod ends, so consult a Moog or TRW catalog for your particular needs.

STEERING

A steering ratio of 16 to 1 is generally considered the ideal ratio for a race car for short track racing. The faster-than-stock ratio helps the driver make small, precise movements with the steering wheel.

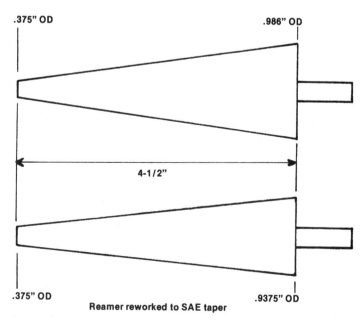

42R No. 2 Reamer

.375" OD .986" OD

4-1/2"

.375" OD .9375" OD

Reamer reworked to SAE taper

The steering ratio is defined as the amount of steering wheel angle movement required to turn the ground wheels one degree. For example, a 16 to 1 ratio means that when the steering wheel is moved 16 degrees, the front wheels are moved one degree. Several factors can influence this ratio — the steering wheel diameter, the pitman arm and idler arm length, the worm and nut assembly in the steering box, and the length of the spindle steering arms. All of these elements must be in harmony with each other to achieve a 16 to 1 steering ratio. It is possible, for example, to use a 16 to 1 worm and nut assembly in the steering box, but have a small diameter steering wheel and very long steering arms, and end up with an overall steering ratio of 24 to 1. Once you have assembled your complete race car, check the combination you have to be sure of the ratio. You can do this by placing the front wheels on alignment plates and affixing an inclinometer to the center of the steering wheel. Synchronize the inclinometer and wheel plates at zero degrees when the tires are straight ahead. Then move the steering wheel until the wheel plates show the front tire has moved two degrees (moving just one degree can cause an error because the steering linkage could be taking up some clearance in the initial movement). Then check the reading on the inclinometer. Divide its degree reading by two, and you have the ratio. For example, if the inclinometer reads 32, divide by 2, and you get 16, which means you have a 16 to 1 ratio.

The most common method of achieving a 16 to 1 ratio is by changing the worm and nut assembly in the steering box. In a Chevrolet Saginaw manual steering box, this is easily accomplished by installing the Corvette part number 5677650 worm and nut.

In the 1965 and later Ford Galaxy steering box, the same Corvette worm and nut can be installed to quicken the ratio, but the Chevrolet intermediate car sector shaft must be substituted in the Ford box for the Ford sector shaft. This will all fit fine, except that the Chevy sector shaft in the Ford box will leave a slightly sloppy fit, and a small amount of grease leakage will occur. If you want to tighten up this fit, take the bearing out of the Ford box and machine a brass bushing to take its place, taking the O.D. and I.D.

measurements from the parts in your modified Ford box.

Slow steering Mustang manual steering boxes can be modified by substituting a Mustang power steering worm and nut assembly for the production worm and nut. The Mustang power steering box ratio is 16 to 1, and it can be used because this car's power steering acts on the drag link and not through the box.

STEERING BOX ADJUSTMENT

It is important to properly adjust the worm bearing preload and sector shaft lash on worm and recirculating ball steering boxes. For the worm bearing preload, tighten the worm adjusting nut until there is a slight drag on the worm shaft when turning the shaft by hand. It should measure between 8 and 10 inch-pounds with an inch-pound torque wrench. Then lock the jam nut in place. Install the sector shaft, making sure that the center tooth is in the center of the worm block. Adjust until there is a definite tight spot when turning the sector shaft past center. The sector shaft preload should work out to about 20 to 24 inch-pounds.

RACK AND PINION STEERING

There are advantages and disadvantages to using rack and pinion steering in a stock car. The advantages are lighter weight, less parts and linkage, ease of adjusting for bump steer, and a built-in quicker steering ratio.

The drawbacks of a rack and pinion unit are its susceptibility to damage, and a tendency to deflect. In the typical steering linkage used on a Chevelle or Nova or other typical American passenger car, the steering system is designed to take loads in bending, and can thus suffer quite a heavy blow at the front wheels with nothing more serious than a bent drag link or tie rod (which can be straightened). However, the rack and pinion steering system takes all of these loads in tension and compression. This means that an impact load will really inflict serious damage to the steering system. A bump to a Chevlele steering linkage might mean a bent linkage piece that would still let the race car finish the race, but with a rack and pinion system it would most probably spell the end of the race for the car (plus mean the installation of a new rack and pinion).

The best American rack and pinion system to adapt to a race car is found in an American Motors Pacer. It is the approximate correct size for most cars, is fairly rugged, and uses GM intermediate car tie rod adjuster sleeves so a variety of tie rod ends can be fitted.

The rack and pinion has to be mounted to a bracket fabricated off the front crossmember of the car. The rack should be located so the arms run parallel to the lower A-arms. The rack can be shimmed up or down, and backward and forward to adjust for any bump steer.

The Corvette worm and nut with bushing and seals to fit the Ford steering box. This kit is available from Speedway Engineering. The Corvette sector shaft must be used with this worm and nut, but it is larger in the spline area than the Ford pitman arm. Have the shaft resplined, or file the pitman arm carefully with a triangular hand file.

FRONT END ALIGNMENT

Any race car's performance is quite critically dependent on proper front end alignment. The car should be tested through skid pad testing, as we have outlined in the Testing section of this book, to find the optimum front end alignment settings for the car. The alignment should be checked weekly before the car goes to the race track, between the heat races and main event at the race track, and definitely after either of the front wheels have taken an impact on the race track.

If you are serious about your race car and its performance, you will purchase a bubble-type caster/camber gauge to set your front end alignment. It is a special tool which costs money, but it is one tool which you will constantly rely on and which is very important to the proper operation of your race car. The best gauge to purchase is manufactured by Bender Equipment Co., and has a scale that goes to plus and minus four degrees. Look in your Yellow Pages under "Wheel Alignment Equipment" to find a dealer, or write to Bender Equipment Co., in South Gate, Calif. or Birmingham, Ala.

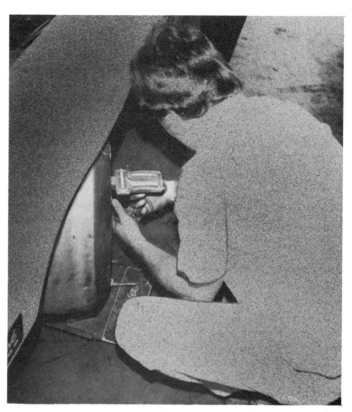

The Bender front end alignment caster/camber gauge is really handy to have, along with their wheel plates. When purchasing the gauge, make sure you get the model that goes up to five degrees negative camber.

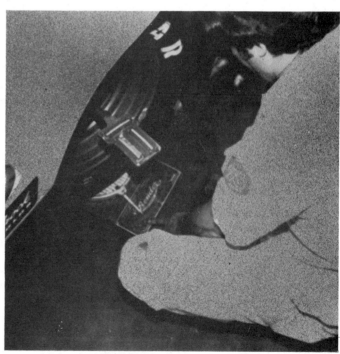

Chassis & Suspension Fabrication & Design

All of the basics of suspension design are fully covered in Steve Smith Autosports' several suspension books, so we won't go into that aspect here. This chapter is designed to lump several chassis building ideas and methods together to put theory into practice, and to present ideas which are only gained through years of experience.

ANTI-DIVE

Anti-dive, which is a force gained by the placement and layout of suspension linkages in the front end of a vehicle, can be achieved in two ways: 1) The upper and lower A-arms are mounted at dissimilar angles as seen from the side view, and 2) The upper and lower A-arms are mounted at dissimilar angles as seen from the top view.

The reason for using any amount of anti-dive is to help eliminate forward pitching of the vehicle during braking. With anti-dive through the linkages controlling some of the front-end dive, softer front spring rates can be used. This aids handling through better tire compliance. The ideal amount of anti-dive for a stock car is 25 percent. Too much anti-dive is not healthy though, as the mechanical linkage advantage of anti-dive is gained by locking up the front suspension mounts. As more anti-dive is added, the suspension mounts lock up more.

When a car is using upwards of 25 percent anti-dive, the driver must be aware of it and the ramifications it presents to him. If the driver's style is harsh, jerky or erratic as he enters a corner, a high amount of anti-dive will come into play instantaneously when the brakes are slammed on, and the car will understeer. If the driver then applies more brake, the car could understeer worse, but if he releases the brakes, the car will handle normal or switch to oversteer if the driver has been expecting and correcting for understeer. It definitely is a situation which can fool a driver if he is unaware of it in his car.

The other problem associated with anti-dive is caster gain. As the percentage of anti-dive increases, so does the amount of front wheel caster gain. With too much caster gain, the wheels can become almost impossible to turn or steer.

ANTI-SQUAT

Anti-squat is a reactive mechanical force very similar to anti-dive, only it is associated with the rear suspension. Anti-squat is a counter reaction of the rear suspension linkages against the vehicle's body squatting under acceleration. Anti-squat acts as a lever between the rear tire contact patches and the body of the vehicle, trying to force the body back up as acceleration is forcing it down. 100 percent anti-squat would mean, then, that just as much effort is being expended pushing the body up as is being expended by the accelerating forces pushing the body down in the rear end. In other words, under 100 percent anti-squat there will be no rearward pitch of the vehicle's body under acceleration.

Anti-squat has two advantages: 1) It helps achieve better

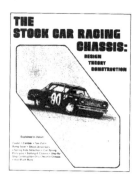

These three suspension books from Steve Smith Autosports cover all the basics of suspension design.

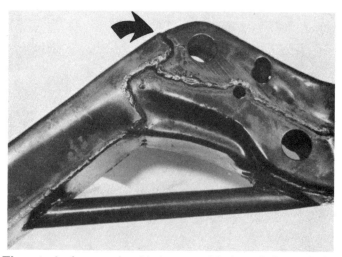

The stock frame should be rewelded and braced for strength. The arrow points to the area of the frame which is notched to lower the car, as discussed in the drawing on page 126.

traction with the rear tires under power, and 2) It helps combat power oversteer. Power oversteer is experienced by a car coming off a turn when the driver feeds more power to the car and the rear wheels break loose and slide outward, rather than gain a forward bite. See the circle of traction explanation in the Tire chapter elsewhere in this book for a more complete explanation.

Just as with anti-dive, there are disadvantages associated with anti-squat as well. First of all, anti-squat must be held to 50 percent or less, or the rear wheels will hop and chatter under braking. And just as with anti-dive, the driver of a car which has a lot of anti-squat in it must be aware of it, or it could get him in trouble. For example, if the driver must lift off the throttle quickly while accelerating off a corner in a car with a great amount of anti-squat, the mechanical force will come off the rear tires real quick and the car will most likely spin.

SHOCK ABSORBERS

The ideal mounting angle for shock absorbers is 5 to 10 degrees away from vertical. This places the shock absorbers in line with the angle of the forces which they will be resisting, without introducing any amount of linkage bind in the shock absorbers. With the shock absorbers angled at any more than 15 degrees away from vertical, the shock will begin to lose its working efficiency. At 15 degrees from vertical, the shock absorber loses 5 percent of its efficiency. At 20 degrees from vertical, it loses 10 percent. At 45 degrees from vertical, it loses 50 percent of its efficiency.

In addition to work efficiency lost, excessive shock absorber mounting angle will also cause oil foaming inside the shock. If the unit is mounted at 20 degrees or more from vertical, there is a great danger of foaming. When the oil foams, you have complete shock fade.

On stock cars racing on an oval track, the left side shock

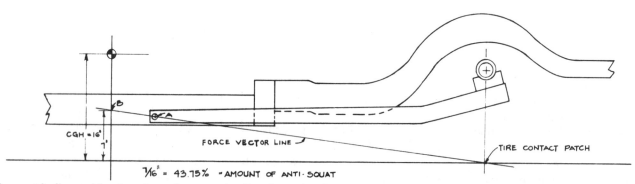

Anti-squat is figured by drawing a force vector line from the tire contact patch through the trailing arm front pivot point [A], and is extended through the lateral center of gravity [B]. If that intersection [B] is 7 inches above ground and the CGH is 16 inches above ground, the intersection of the force vector lies 7/16 or 43.75% of the distance from the ground to the CGH. This, then, is the amount of anti-squat.

This is a square tube frame which butts up to a Camaro front stub frame. Note how the joint is made, and how cross bracing with round tubing adds strength.

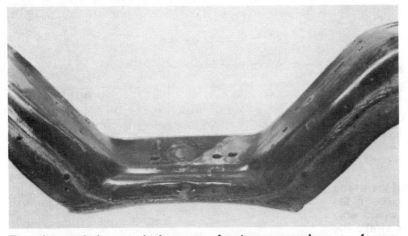

To gain needed ground clearance, front crossmembers on frames such as a Chevelle need to be cut and rewelded such as this.

Note the sturdy triangulated engine mount, and the three-point reinforcement of the joint of the Grand National-type front end to the tubular frame rails.

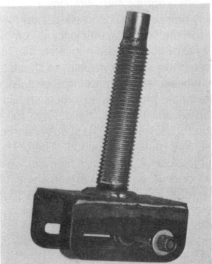

Weight jackers are simple to construct. Top, a 3/8-inch plate is flame-cut into a round shape and rod material is bent and welded in place to stabilize jacker over spring. Then a section of 1-inch O.D. All-Tread such as seen above is obtained and a hole smaller than All-Thread O.D. is drilled in plate. Bottom portion of All-Thread is machined smaller so there is a shoulder step which bears on plate, then placed through hole in plate. A hole is drilled through protruding shank on All-Thread below the plate and cotter pin is placed through it to retain screw in plate. Large nut is placed on All-Thread as seen below and is welded to spring bucket or frame.

absorbers should be mounted vertical, or as vertical as practical. Even a slight amount of positive camber angle on them would not hurt. This should be done to prevent centrifugal force from having such a great affect on the oil in the shock absorber. If the left side shock absorbers are angled over at the top towards the center of the car, the angle will simply aid the centrifugal force in pushing the oil in the stock up to the top and to one side.

Tie rod end shock absorbers should be mounted so that the ball rotates with the suspension arm movement, and not so that the suspension arm simply moves the tapered stud of the tie rod end. The bottom of the shock absorber should also be mounted as close to the tire as possible to be sure that all small bumps and vibrations are controlled.

Many times shock absorbers get bottomed out or simply worn out, their working efficiency is completely lost, yet the car owner or builder does not catch that fact. That amounts to running the car with virtually no shock absorber at one or more wheels. As a matter of regular weekly race car maintenance, the shock absorbers should be removed and stroked several times by hand. Two full cycles should be enough for each shock. The shock absorber should feel firm all the way in each direction. There should be no easy resistance. Let us emphasize here that you are checking for broken or severely worn parts in the shock absorber with this exercise. You can never feel a true shock absorber force by hand stroking it. There is a small leak built into the unit so it can be stroked by hand for ease of installation. It takes roughly 400 pounds of force to open the damping valves on a stock car shock absorber.

At right, the two drawings indicate the right and the wrong ways to mount the shock absorbers to the chassis. In the wrong way, small vibrations and deflections would only work the tie rod end up and down, and never stroke the shock absorber. Below, the drawing shows the maximum angle at which to mount shock absorbers in the rear end. The width of measurement E should be as close to measurement F as possible to get the most efficiency from the shocks.

The rubber shock absorber travel indicators found on the stem of shocks can be useful to you in setting up the chassis, and determining what the chassis is doing. Once you have your car all set up, take the springs out of the right front and right rear (one at a time) and jack each wheel through ½-inch increments of wheel travel. Note the amount of shock absorber travel versus the amount of wheel travel, and make a chart of the results. That way you can measure shock travel at the track and relate it directly to wheel travel and spring travel.

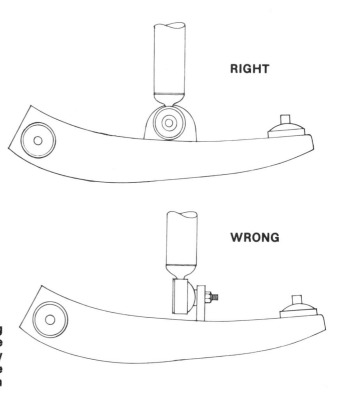

RIGHT

WRONG

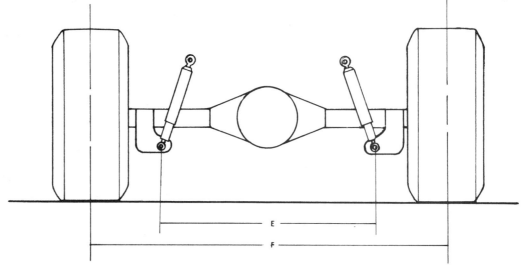

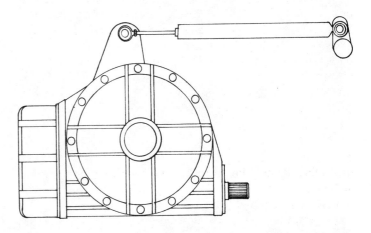

Above, Carrera makes a special shock absorber called an axle dampner which positively controls leaf spring wrap-up. Left, many people place lengths of chain from the housing to the frame to serve as a maximum bump stop so spring will not fall out of pocket when car is jacked up. Below left, run the wheels through maximum travel to determine how much body roll clearance there is. Any chassis should have at least 5 inches of bump travel available.

ENGINE LOWERING

There have been an increasing number of race cars showing up at the tracks lately with the engine lowered to a height of five inches measured from the ground to the crankshaft centerline in the front. This engine lowering significantly lowers the center of gravity height, yielding better handling especially in entering corners. It also entails a lot of work and a large expenditure of money (remember, speed costs money!).

The first thing required is a shallow dry sump oil pan, and naturally a complete dry sump system. The externally mounted dry sump pump must be moved up above the crankshaft centerline to allow ground clearance, and prevent damage to the pump and belts.

The next thing in the front of the engine which gets in the way is the crankshaft dampner. On a Chevy small block engine, this is easily rectified by swapping the stock 350 dampner for a Chevy 409 dampner. It is smaller in diameter than the 350 piece, but it has the same mass (be sure to have the 409 dampner balanced with your crank, flywheel and clutch rather than just bolting it in place on an existing engine).

Going to the rear of the engine, a stock-sized Chevy small block flywheel and clutch assembly is too large in diameter. Use a Fiat 850 flywheel in conjunction with a Borg and Beck triple plate clutch, which will get the size down for proper clearance and considerably lessen the rotating mass. Use the Chrysler gear drive starter with a Quarter Master Industries adapter kit. An alternate choice for the Fiat 850 flywheel is the flywheel and 3-horsepower gear drive starter from a Mazda rotary engine.

The final clearance obstacle is the transmission, with its bulge for the cluster gears protruding too low. To remedy this, the bellhousing must be redrilled in order to relocate the transmission bolt pattern with the transmission rotated 45 degrees counter clockwise. The transmission shifter will then mount to a plate located on the chassis, and the linkage will run to the transmission as normal.

Headers must be used which curve above the engine and run over it, exiting at the right rear of the engine compartment.

ANTI-ROLL BARS

Special anti-roll bars for race cars can be purchased in a

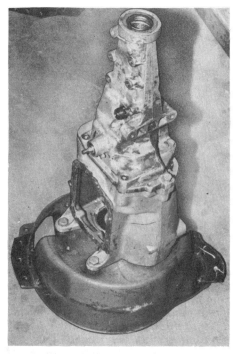

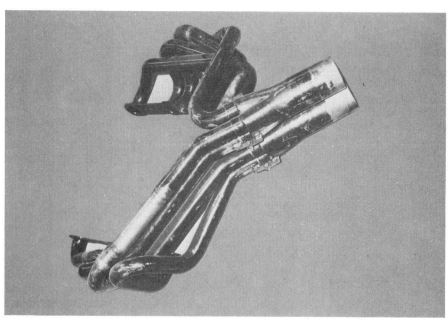

To get the engine in the chassis with the crankshaft centerline to ground measurement at 5 inches, the transmission must be turned 45 degrees [above left] and headers must be used which exit above the engine [above].

variety of diameters and lengths from several firms specializing in the manufacture of race car chassis parts. Expect the bar, arms and mounting kit to cost upwards of $100. There is a very workable and financially sound alternative, however, for the racer on a budget.

There are a variety of heavy duty passenger cars and trucks being made today which are equipped with anti-roll bars which are heavy enough to fill the needs required. The bars also are found in a variety of sizes so that they can be adapted to most any type of chassis and to fit most spring rate requirements.

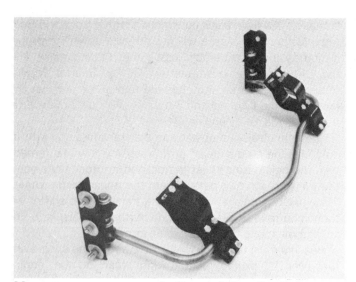

Many passenger car and aftermarket anti-roll bars are available with an appropriate bar diameter, but try to stay away from bars which do not run straight. Rating them is almost impossible.

This is the Firbird Trans Am 1.25-inch diameter bar in the stock rubber mount, employed on a race car. Very effective bar.

This is an aluminum pillow block which can be machined to use as an anti-roll bar mount. Clearance between bar and mount should be no more than .015-inch. The bolts which hold the pillow block to the base plate are welded to the plate. Be sure to use a bolt like a Grade 3 so the welding does not cause brittleness in the bolt.

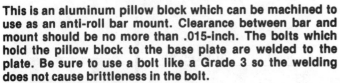

Aluminum pillow block is attached to roll cage tube [very upper right of photo] above the frame and arm and link attach to upper A-arm. Very efficient system. Note that bar is tubular — good weight saving idea. A shaft collar with set screw must be used on the bar on either side of pillow block. Below left and right, the pillow block mount is used for a rear anti-roll bar.

In the accompanying chart, we have listed a variety of anti-roll bars which we have measured and calculated. You can see that there are several bars in the chart which have a spring rate much greater than many of the commercially available racing bars. When using the stiffer bars, you must be sure that the mounts attaching the bar to the chassis are absolutely rigid. A sloppy or loose mount will cancel out some of the spring rate effectiveness of the bar by allowing the mount to deflect instead of the bar.

Once an anti-roll bar is chosen and mounted, the bar and its linkage must be checked for bind and interference. Be sure the tires do not contact the bar arm or connecting linkage when the wheels are turned at full lock in both directions, and when the wheels are at full droop or full bump. Also be sure that any spherical joints or bushings in the connecting linkage do not bind up through the full normal travel of the wheels in bump and rebound. Any type of bind will cause broken linkages or mounts.

The most effective type of mounting bracket and bushing for an anti-roll bar is a fabricated two-piece aluminum pillow block (see photo). A machine shop can easily make this for you from a piece of aluminum bar stock. Also, any of the Corvette or Firebird mounting hardware can be used for the Corvette or Firebird anti-roll bars we have specified in the chart should you have difficulty in machining the aluminum pillow block mounts. The only drawback to passenger car-type anti-roll bar mounts is that they employ rubber bushings which deflect, and consequently work loose very quickly. You will find that you will be purchasing quite a number of replacement bushings for the mount during the season should you decide to use stock mounts.

SPRINGS

It is important to rate each coil spring which you purchase, whether it is new or used. New springs are subject to manufacturing tolerances as well as mislabeling. Don't assume anything. Used springs are subject to internal stresses, which can weaken them.

If you have a set of springs which you are using in your race car week after week, do not assume that they will keep their rate. Coil spring wire is under high stress, and it weakens. At least once every other week take the springs out and re-rate them to be sure of their true rate. You may be surprised to see the change in rates.

Production Anti-Roll Bars

Vehicle	Part #	Rate	OD	Bar Length	Arm Length
Corvette	351596	2000	1.125	33	5¼
Corvette	3831972	968	.9375	33	5¼
Corvette	3871318	726	.875	33	5¼
Corvette	334930	548	.8125	33	5¼
Chevy truck	Blazer	531	1.25	23	15
Chevy truck	Blazer	838	1.25	23	12
Chevy truck	¾ -ton pick-up	653	1.25	29½	12
Chevy truck	¾ -ton pick-up	931	1.25	29¼	10
Firebird	3984557	247	1.0	34	11½
Firebird	3975523	395	1.125	34	11½
Firebird	3986480	604	1.25	34	11½

This is a spring rate checker available from Steve Smith Auto-sports.

Note the compound mounting angle of the shock absorber. It should be avoided if possible, because the angular movement decreases efficiency of shock.

Tubular anti-roll bars such as these are available in a variety of sizes from Speedway Engineering.

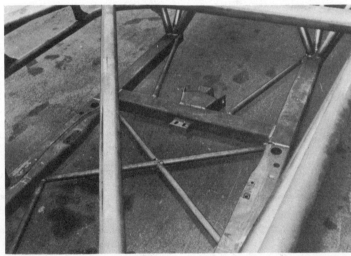

These four photos illustrate a very well designed and triangulated chassis. There are only four bends and one radius in the entire chassis exept for door bars. Good for ultimate strength, and very easy to construct. Forget the old idea that a good roll cage must have all kinds of bends.

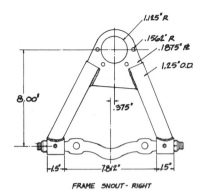

FRAME SNOUT - RIGHT

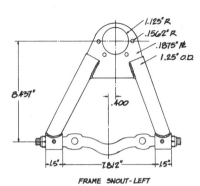

FRAME SNOUT - LEFT

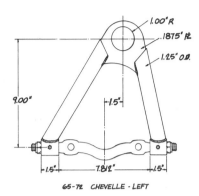

65-72 CHEVELLE - LEFT

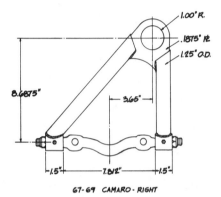

67-69 CAMARO - RIGHT

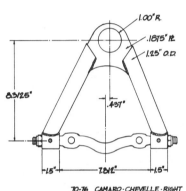

70-76 CAMARO - CHEVELLE - RIGHT

Blueprints for a variety of upper A-arms which you may need to construct. Inner control shaft is from a 1965 and later Ford Galaxy. Tubes are made of 1.25-inch O.D., .095-inch wall tubing. Plate which holds ball joint is flame-cut from .25-inch thick steel.

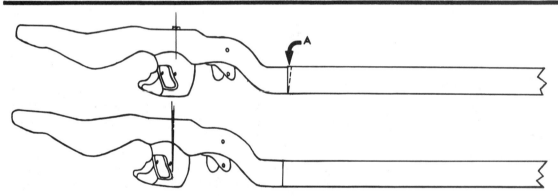

These two drawings illustrate a virtually undetectable method of lowering a car with a stock frame without any frame "altering." A V-cut is made at A in frame, then front portion of frame is pulled up and rewelded. The very top of the V-cut should be 5/16-inch wide [where arrow points], and this will result in the chassis being 1-inch lower as measured from ground to the top of weight jacker. A similar alteration should be followed at the rear of the frame rails where the kick-ups join.

The lower control arm in your chassis should be checked periodically for cracks just in back of ball joint hole. They are very common here. Lower A-arms receive 3 times the forces that the upper A-arms do.

Wheels

DIMENSIONING

The important dimensions for properly choosing racing wheels is rim diameter, rim width, bolt pattern and offset.

The rim diameter is measured from the bead seat across the center to the bead seat. It is not the overall diameter measured from the outside flanges.

The rim width is the wheel width measured, again, from bead seat to bead seat, and not flange to flange. The selection of the proper rim width is tied directly to the tread width of the tire to be used. When a tire is mounted on too narrow a rim, it develops a protruding crown in the center of the thread surface. Wider wheel rims will allow a greater cornering power than narrow ones because they do not let the tire sidewall become rounded (meaning less sidewall movement), and they allow a flatter tire contact patch with more tire area on the ground. The best rule of thumb is to use the widest rim possible.

The bolt pattern refers to the number of lug bolts the wheel uses as well as the diameter of a circle drawn through the centers of the lug bolt holes. The standard stock car racing bolt pattern is 5x5. This means that the wheel employs five lug bolts, and the bolt mounting holes are drilled on a circle diameter of five inches.

Offset is the lateral distance from the hub mounting surface of the wheel to the centerline of the wheel. Positive offset places the centerline of the wheel toward the centerline of the car, while negative offset places it toward the outside of the car.

Wheel offset has a great deal of interaction with the tire scrub radius, which in turn has an influence on handling characteristics. The scrub radius is drawn from the point where the steering axis line intersects the ground to the centerline of the tire contact patch. The greater this distance, the more the tire is pulled or scrubbed in an arc around the steering axis intersection point. Wheel offset can alter the severity of the scrub radius. Generally, positive wheel offset will decrease the scrub radius, while negative offset will increase it.

Positive wheel offset, because of the influence of the scrub radius, promotes stability under braking and ease of steering. Negative offset, on the other hand, creates hard steering and places a heavy load on susension components such as wheel bearings and spindles. But negative offset also widens the tread width which creates better cornering ability through lessening of weight transfer during cornering.

Wheel offset usually ends up as a compromise between ease of steering and providing enough clearance for suspension components and brakes.

WHEEL MATERIAL SELECTION

In many classes of racing, the rules limit wheel selection to steel wheels. In this case, there are two critical factors to weigh: weight of the wheel, and wheel stiffness.

Weight-wise, there are now several major steel racing wheel manufacturers competing to produce the lightest

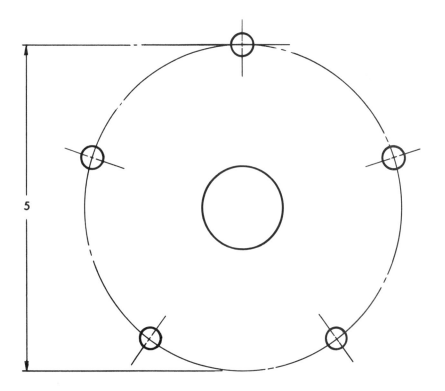

The standard of the racing industry for wheel bolt patterns is 5x5, which means five bolt holes evenly spaced on a 5-inch diameter circle. If you happen to use brake drums or hubs which have something other than a 5x5 pattern, a machine shop can easily redrill the new bolt pattern for you. More often than not, even with a passenger car 5x5 pattern, holes will have to be redrilled to accommodate racing-sized wheel studs.

possible steel racing wheel as racing comptitors become more aware of the negative effect of unsprung weight.

Wheel stiffness can also play an important part in handling characteristics. There are heavy duty production-type steel wheels which have been measured to deflect enough to cause as much as one degree of camber change. This introduces a great unknown into thc handling characteristics of the chassis.

For racing purposes, steel wheels have two advantages: they are relatively inexpensive, and they are durable. They can endure a great deal of rough treatment as well as long service without cracking or breaking. Usually the first area to crack on steel wheels is the welded joint between the center section and the rim, but if detected early this can almost always be easily repaired by welding.

Lightweight aluminum alloy or magnesium wheels, on the other hand, are not nearly as resistant to rough useage. And when they crack or are damaged in an accident, they must be discarded.

Present day all-alloy wheels are almost universally made from 356-T6 heat treated aluminum alloy, rather than magnesium. The porosity of the aluminum alloy is less than magnesium, the cost is less, and the strength is greater. In fact, a good quality alloy wheel will be superior to a steel wheel in terms of strength and stiffness.

The quality — and thus the price and serviceability — of alloy wheels is dictated by the forming operation of the wheel. They may be cast or forged. The forged wheel will be

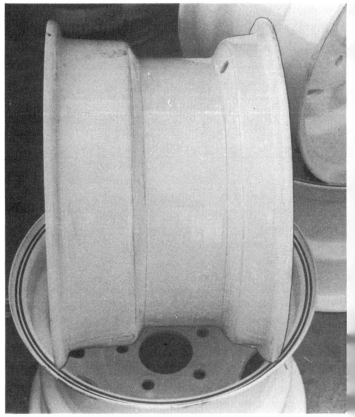

This is the Norris Industries Grand National racing wheel, which is designed to accommodate any size disc brake caliper.

stronger, and more expensive.

Any alloy wheel used on a stock car should have steel seats for the lug nuts to snug up to. This will prevent nuts gouging into the wheel material, and prevent lug nut loosening.

WHEEL MAINTENANCE

All newly purchased wheels should be checked for lateral and radial run-out. Lateral run-out is sideways movement, commonly referred to as wobble. Radial run-out is the true roundness of the wheel. Check both dimensions with a dial indicator gauge. If either run-out is more than .025-inch, discard the wheel and return it for another.

After each race, all wheels should be carefully checked for cracks and dents. Any cracks or gouges in an alloy aluminum wheel means it is time to replace that wheel. On a steel wheel, a crack in the rim-to-center section can be welded, but all other cracks and dents is good cause to replace the wheel. Wheel cracks can grow quite rapidly when subjected to high speed and high side loadings, so it is not sensible to risk the safety of the driver and car with a faulty wheel.

BALANCING

All racing wheels and tires are subjected to high speeds, so balancing is definitely required to keep unwanted vibrations out of the chassis and to promote better tire compliance with the track. There are three methods of balancing (static, dynamic mounted and dynamic unmounted), but the accuracy of any balancing method is not nearly as dependent on the method itself as it is on the skill and care of the person performing the balancing.

Static balancing requires the balancing of the tire and wheel on a bubble balancer which shows an excess of weight in one direction. This method, if performed correctly, gives accurate balance of the wheel and tire assembly by itself. This will allow this tire and wheel assembly to be used at any one of four positions on the car without changing its balance.

Mounted dynamic balancing, on the other hand, balances a particular wheel and tire assembly along with the brake drum or disc at the one corner of the car where it is mounted. This can present a disadvantage in moving a dynamically balanced tire and wheel to another position on the car.

Unmounted dynamic balancing is the best method of the three, but it requires a floor-mounted dynamic balancer which is not readily available to many people. The dynamic balancing of the wheel and tire off the car can cure many imbalances in the assembly which are not curable with static balancing.

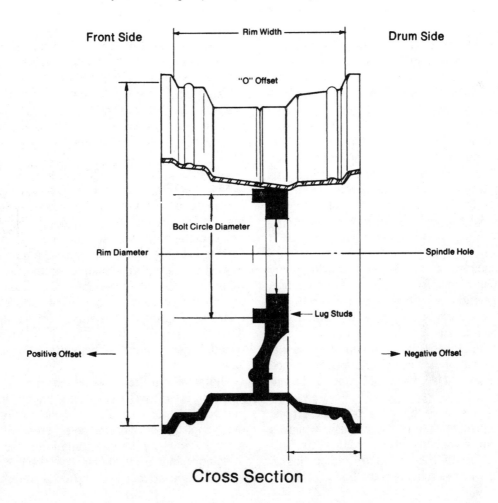

Cross Section

Tires

Tires are the only link a race car has between the vehicle and the track. This means that the tire selection must be correct in order for good handling and driver skill to have an effect on the performance of the car. A poor choice of tires will make a good car uncompetitive. So, in order to receive the ultimate in performance from the car and suspension, a tire must be selected which will perform best for the particular situation. And to get the most from that tire, the chassis should be designed from the tires up. What this means is knowing what makes the tires work best. For example, what downward forces can the tire support? What lateral loads can it support? At what inflation pressure? At what temperatures? What negative camber angle does it require?

Because of the complexity of racing tires and the many thousands of tires being manufactured by many companies (Firestone Racing Division alone makes over 600 different racing tires), recommending a particular tire brand and model for any certain useage is not possible in this book. But what is intended here is a broad educational background on which design factors effect tire performance in certain situations to enable you to pick tires which are best for your needs.

First of all, let's define all of the terms peculiar to tires. The basic framework of the tire is the carcass, which is made up of various fabrics and materials piled in layers. This carcass is then covered with rubber to form the outside of the tire. The thickness of this rubber on the tread surface is known as the skid gauge. The thickest skid gauge used on racing tires is usually 7/32-inch, and the thinnest 3/32-inch. The average new racing tire has 5/32-inch.

Lateral spring rate of a tire is the technical description of the sidewall stiffness. The profile is the total heighth of the tire, whereas the aspect ratio is the ratio of the sidewall heighth to the tread width.

TIRE SIZE

Now for some generalities about racing tires, and their explanations. The wider the tire, the better it works. There are two reasons for this: first, there is simply more rubber present to grip the track surface, and secondy, the more tire surface in contact with the track, the more surface there is to absorb wear. Consequently, as the tire gets wider, the rubber compound can get softer. To imagine why this is so, picture the process of scraping a rubber eraser across a metal file. The smaller the eraser surface, the faster it wears down. But a wide surface does not wear down nearly as fast because there is more rubber surface present to absorb the wear.

The smaller a tire's overall diameter is, the smaller the footprint area it will have to put on the track surface. To compensate for lost footprint area, tire engineers are designing the smaller diameter tires with more flexible sidewalls so more tire will come in contact with the track. Judging from this knowledge, it is wise to choose a tire with a taller carcass when running on a track with a rough or

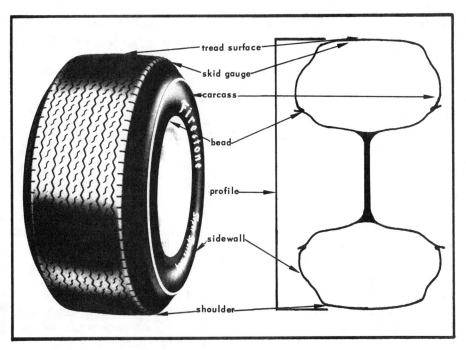

The basic terminology associated with tire construction.

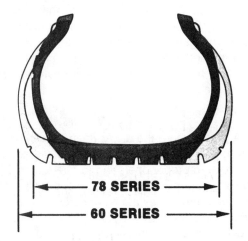
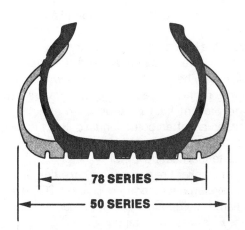

This shows relative differences in aspect ratios.

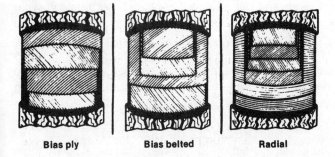

Bias ply **Bias belted** **Radial**

The three basic types of tire construction

bumpy surface so the stiffer sidewall will help cope with the bumps and irregularities.

The more rubber, the better the bite. That is the explanation of why slicks work best on short tracks, and why the superspeedway tires are for all practical purposes a slick tread design. The tread on tires is there mainly to provide an escape for road material such as dust, small particles of dirt, etc., which get trapped under the face of the tire.

RUBBER COMPOUNDS

It is best to use the softest possible rubber compound, but there are several limiting factors to this. Soft compounds work better because they conform to the road better, thus they act like adhesive tape-wrapped wheels running on a slick track. But on a porous surface a soft compound tire gets worn down like it is in a lathe. When a tire's compound is too soft for the track surface, the tire will develop a "grain" pattern across it which is caused by the track actually trying to rip away the surface of the tire.

One more factor about soft tire compounds. They grow in overall diameter during operation because their rubber is more pliable.

If your racing association should pass a rule limiting your class of racing to one hard compound tire, you can decrease the rate of the springs on the outside of your chassis to gain more body roll and thus more downforce on the tire to produce a better tire bite. On a given asphalt track with little porosity, a hard compound tire will not bite nearly as well as a soft compound tire, and thus will require more downforce on the tire to gain more traction.

Weather conditions (ambient, or outside, temperature and the amount of sunlight) will influence the choice of tire compound at any track. For example, a cool outside temperature or an overcast day, will dictate a softer compound.

To increase tire life, and get more mileage from tires, choose tires with both a harder compound and an increased skid gauge.

To prevent tire blistering, choose a tire with a thinner skid gauge, a taller overall diameter, and a wider tread. To prevent blistering, be sure to scuff in a new tire and allow it to cool before running a race on it. Change tire compound hardness to prevent heat build-up from effecting the tire (choose either a softer compound so the tire wears out before heat build-up occurs, or a harder compound that will sustain all potential heat build-up).

To select a quicker tire compound, take tire temperatures and if the average temperature is over 260 degrees F., choose a harder compound. If the temperature is less than 200 degrees F., choose a softer compound.

INFLATION PRESSURE

To increase traction, run the inflation pressure as low as possible without creating instability or sucking the bead off the rim. The correct inflation pressure can be found by checking tire temperatures. Another way of gaining better traction is to put the tire on the car against the direction of rotation the arrow on the tire shows. This will run the rubber against the grain. A tire will not last quite as long running like this because the overlap will start to separate.

In general, a change of ten pounds of inflation pressure will change the overall diameter of the tire by .2-inch, change the cross section of it by .025-inch, and change the shoulder drop by .05-inch.

In general, a change of one inch in rim width will change the cross section by .3-inch, change the overall diameter by .25-inch, and change the shoulder drop by .1-inch.

The shoulder drop is encountered when the shoulder of the tire becomes lower than the center of the tread surface

when mounted and inflated (looking at the tire from a cross section view). For example, when a tire is mounted on too narrow a rim, it develops a protruding crown in the center of the tread surface. This is because the shoulder of the tire has dropped lower than the tread surface.

THE SIDEWALL

The sidewall of the tire is built so that it will flex. If it did not, when you cornered hard the rim would move outward in relationship to the footprint, and the rim movement would pick up the inside shoulder, lifting the tread surface right off the track.

Wider wheel rims allow a greater cornering power because they do not let the sidewall become rounded (thus less sidewall movement) and they allow a flatter contact patch with more tire area on the ground.

INNER LINERS

When using tires with inner liners in them, there should always be a 20 PSI inflation pressure difference between the outside tire and the inner liner (the inner tire should have the greater tire pressure). The inner liner pressure should be kept as high as possible so that in case of an outer tire blowout, there is not a great distance for the wheel to drop onto the inner liner. But too high an inner liner pressure will cause it to bulge against the tire carcass and heat up the outer tire.

TIRE WARMING

A place where the driver must be certain to take care of his tires is just before a race when he should be bringing them up to proper operating temperatures. Tires will exhibit all sorts of unpredictable characteristics if too much is asked of them before they are warmed up. The best way to warm tires during the pace laps is to zig-zag the car and use up as much track as possible.

CAMBER

As mentioned in the beginning of this chapter, every tire has certain specifications of what downforce and lateral loads it can withstand, at what ideal inflation, temperature and negative camber angle. All of these characteristics are intertwined and one effects the next. Unfortunately, what these characteristics are for each tire only a select few top racing teams are ever allowed to know. But one of these facts you should be aware of is that every racing tire needs at least a half a degree of negative camber under full load to corner properly and develop its full potential cornering power.

THE CIRCLE OF TRACTION

The circle of traction theory helps to relate down force loadings to traction and forward bite. The total amount of traction forces available at a tire can be expressed in pounds of force. This force is available for either side bite, forward traction or both. The traction forces must be reacted by down loadings of weight on the tire. For example, if one pound of downward weight is placed on a tire, one pound of traction force can be expected from it. But, that traction can only be one pound's worth in one direction, or else split between a half a pound of side bite force and half a pound of forward traction force. If 1000 pounds of traction force are available to the right rear tire, and 600 pounds of force are required for maximum side bite of that particular tire for that particular load, then only 400 pounds of force are available for accelerating. If the throttle was opened up all the way and more than 400 pounds of accelerating force was applied to the tire, the forces would overload the tire and remove side bite ability. The result would power oversteer, or the rear end of the vehicle oversteering from too much throttle application. This problem can be overcome by feathering the throttle to the maximum traction ability of the tire, or switching to a different tire that has a softer compound, greater overall diameter or wider tread width. Softer outside spring rates could also help to apply more downforce.

RACING WITH PASSENGER CAR TIRES

Many racing association rules call for racing with passenger car tires in order to cut costs for the competitors. Several guidelines can be followed in order to pick the tires which will give the most performance and cornering power.

Start by choosing a tire with a 4-ply nylon carcass. Many different fabrics — all of which are in use in passenger car tires — have been tried in racing tires. All of them have exhibited problems so far, except for the nylon fabric which racing tires are currently made from. Polyglass cords have a problem of breaking at racing speeds. Rayon does not have the tensile strength required for racing. Steel cords and polyglass cords both have problems of tread separation at racing speeds. Steel radials are also too heavy. The only other tire cord material to use other than the nylon is Goodyear's Flexten, which is an aramid fiber used in its Customgard GT Radial tires. It is good for racing use, and in fact, was used in racing in IMSA's GT Radial Series.

The accompanying chart lists many different tires which are available for use as a racing tire. All of these were chosen for inclusion in the chart because they have characteristics very similar to racing tires.

You will notice in the chart that only 15-inch wheel diameter tires were chosen. This is because a larger overall diameter tire is available with the larger wheel diameter, and the larger overall diameter is more desirable.

Tire tread compound hardness can be checked from tire to tire with the aid of a durometer. Be sure to do this to get the softest possible compound.

Use a tire truing machine to shave the tread depth on passenger car tires used for racing. The normal new passenger tire has 11/32-inch of rubber. This should be shaved down to a maximum of 5/32-inch. Shave one new set down to 3/32-inch for use as a qualifying set. The tread is shaved to prevent heat build-up and handling instability from tread squirm.

Be sure to choose as wide a tire as the racing rules will allow, and use as wide a rim as allowed.

GETTING THE MOST FROM YOUR TIRES

To get the most in performance from your tires, the suspension must be adjusted for the proper camber and toe-out. The tires must also be precisely inflated to the correct pressures. These factors not only affect tire performance, but tire wear as well.

The **only** way to precisely monitor the operating performance of your tires is with a tire pyrometer. It will tell you for example if the inflation pressure is too high and the center of the tire is wearing out, or if the right front camber is too far in the negative direction, wearing out the outside edge of the tire. All of these components combine to give you either good or bad handling performance and tire wear.

When new tires are first installed and run a few test laps, the tire pyrometer can perform a vital function for you. It will tell you if the front end alignment is off, or the inflation is incorrect. At that point, you can make the necessary adjustments and be assured of getting maximum life from your tires. However, if a race or two is run on a tire with the inflation too great or camber wrong, spot wear patterns will have alredy developed and no corrections will save the tire life or performance.

Above, the tire pyrometer available from Steve Smith Autosports. Left, Firestone'a Super Sport street tires which can be used for racing where a street tire rule is in effect.

STREET TIRE SUGGESTIONS

Brand	Model	Size	Tread Width	Overall Diameter	Construction
Micky Thompson	Indy Profile	C60-15	6-3/4	24-1/4	4-ply nylon
Micky Thompson	Indy Profile	E60-15	8	25-3/4	4-ply nylon
Micky Thompson	Indy Profile	G60-15	8-1/2	26-1/2	4-ply nylon
Micky Thompson	Indy Profile	J60-15	9-1/2	27-1/2	4-ply nylon
Micky Thompson	Indy Profile	L60-15	10-1/2	28	4-ply nylon
Micky Thompson	Indy Profile	H70-15	7-1/2	28-1/2	4-ply nylon
Micky Thompson	Indy Profile	J70-15	8-1/2	28-3/4	4-ply nylon
Micky Thompson	Indy Profile	L70-15	9	28-3/4	4-ply nylon
Micky Thompson	Racing Profile	E60-15	8	25-3/4	2 ply polyester 2 ply fiberglass belts
Micky Thompson	Racing Profile	G60-15	8-1/2	26-1/2	2-ply polyester 2-ply fiberglass belts
Micky Thompson	Racing Profile	J60-15	9-1/2	27-1/2	2-ply polyester 2-ply fiberglass belts
Micky Thompson	Racing Profile	L60-15	10-1/2	28	2-ply polyester 2-ply fiberglass belts
Concorde	Trac-Action 50	E50-15	8-3/4	24-1/2	4-ply nylon
Concorde	Trac-Action 50	G50-15	9-1/2	25-1/2	4-ply nylon
Concorde	Trac-Action 50	L50-15	9-1/2	26-3/4	4-ply nylon
Concorde	Trac-Action 60	E60-15	7-1/2	25-7/8	2-ply polyester 2-ply fiberglass belts
Concorde	Trac-Action 60	F60-15	8	26-1/3	2-ply polyester 2-ply fiberglass belts
Concorde	Trac-Action 60	G60-15	8-1/2	26-5/8	2-ply polyester 2-ply fiberglass belts
Concorde	Trac-Action 60	J60-15	9	27-1/2	2-ply polyester 2-ply fiberglass belts
Concorde	Trac-Action 60	J60-15	9	27-1/2	2-ply polyester 2-ply fiberglass belts
Concorde	Trac-Action 70	H70-15	6 7/8	27-3/4	2-ply polyester 2-ply fiberglass belts
Concorde	Trac-Action 70	G70-15	6-5/8	27-3/4	4-ply polyester
Concorde	Trac-Action 70	H70-15	7-1/8	28-3/8	4-ply polyester
Pro-Trac	Racing Profile	E50-15	9-3/4	24-5/8	4-ply nylon
Pro-Trac	Racing Profile	N50-15	12-3/4	27-1/2	4-ply nylon
Pro-Trac	Racing Profile	E60-15	8	25-1/2	2-ply fiberglass belts 2-ply polyester
Pro-Trac	Racing Profile	G60-15	9-1/3	26-1/2	2-ply fiberglass belts 2-ply polyester
Pro-Trac	Racing Profile	L60-15	10-1/2	27-7/8	2-ply fiberglass belts 2-ply polyester
Pro-Trac	Racing Profile	G70-15	8-5/8	27-1/2	2-ply fiberglass belts 2-ply polyester
B. F. Goodrich	Radial T/A 50	GR50-15	9	25-1/2	2-ply rayon 4-ply Dynacor belt
B. F. Goodrich	Radial T/A 50	LR50-15	10	26-3/4	2-ply rayon 4-ply Dynacor belt

B. F. Goodrich	Radial T/A60	FR60-15	7-5/8	26-1/3	2-ply rayon 4-ply Dynacor belt
B. F. Goodrich	Radial T/A 60	GR60-15	8	26-7/8	2-ply rayon 4-ply Dynacor belt
B. F. Goodrich	Radial T/A 60	HR60-15	8-1/16	27-1/2	2-ply rayon 4-ply Dynacor belt
B. F. Goodrich	Radial T/A 60	LR60-15	8-7/8	28-1/3	2-ply rayon 4-ply Dynacor belt
B. F. Goodrich	Belted T/A 60	F60-15	8	25-7/8	2-ply fiberglass belts 2-ply polyester
B. F. Goodrich	Belted T/A 60	G60-15	8-3/8	26-3/8	2-ply fiberglass belts 2-ply polyester
B. F. Goodrich	Belted T/A 60	L60-15	9-3/8	28	2-ply fiberglass belts 2-ply polyester
B. F. Goodrich	Belted T/A 70	G70-15	6-1/2	27-1/2	2-ply fiberglass belts 2-ply polyester
B. F. Goodrich	Belted T/A 70	H70-15	6-7/8	28-1/8	2-ply fiberglass belts 2-ply polyester
Goodyear	Customgard GT	F60-15	9-1/8	26-1/4	4-ply Flexten
Goodyear	Customgard GT	G60-15	9-1/2	27	4-ply Flexten
Goodyear	Customgard GT	H60-15	9-3/4	27-5/8	4-ply Flexten
Goodyear	Customgard GT	G70-15	6-1/2	27-1/2	4-ply Flexten
Goodyear	Customgard GT	H70-15	6-7/8	28-1/8	4-ply Flexten

Three of the tires discussed in the table are [from left to right] the B. F. Goodrich Radial T/A 50, the Concorde Trac-Action 60, and the Mickey thompson Racing Profile.

The Clutch

Before getting into a discussion of picking a clutch and installing it, it is a wise practice to define some of the clutch operational basics and the basic styles.

CLUTCH BASICS

Pressure plate — This is the pressure source which keeps the clutch engaged.

Clutch disc — is the member which links the pressure plate to the flywheel.

Flywheel — provides the friction surface which the disc is thrust against.

Holding pressure — This is the pressure which pushes the pressure plate, clutch disc and flywheel together. The holding pressure comes from two sources — static spring pressure, and centrifugal force pressure which is generated by weights.

Static pressure — is the compressed spring holding pressure. The static pressure is not dependent on engine RPM's to generate centrifugal force, but only on spring pressure built into the clutch statically.

Installed spring pressure — is the force exerted based on the installed heighth of the springs between the pressure plate cover and the pressure plate.

Spring compression — The more the springs in the pressure plate are compressed, the greater the pressure exerted. The disc thickness can control spring compression — the thicker the disc, the closer to the clutch cover it is, meaning more spring compression. Increased spring pressure means greater loads placed against the crankshaft thrust bearing when the clutch is disengaged. This is tough on thrust bearing wear. This is why you should always start the engine in neutral with the clutch disengaged.

Centrifugal pressure — Centrifugal pressure is exerted by devices which change their position in the clutch, depending on engine RPM's. These devices are called either centrifugal force assist rollers (on Borg and Beck style) or counterweighted clutch release levers on long style.

Centrifugal assist rollers — are rollers located in slots between the clutch cover and the pressure ring. As the engine RPM's increase, the rollers move toward the outside of the clutch cover. The clutch cover is bolted solid to the flywheel, so the rollers push against the pressure ring, adding to the holding force.

Counterweighted clutch release levers — are levers mounted in a circle to a mount on the pressure ring. The levers are weighted on their ends near the outside of the clutch cover diameter. As the RPM's increase, the weights pivot, and increase the holding pressure. Large counterweights are undesirable for race cars because it makes shifting at high RPM's almost impossible.

CLUTCH STYLES

Most all of the major clutch component manufacturers offer three basic styles of clutches. They are:

Borg and Beck — This style of clutch can feature 0, 3 or 6 centrifugal assist rollers to add to the coupling pressure in

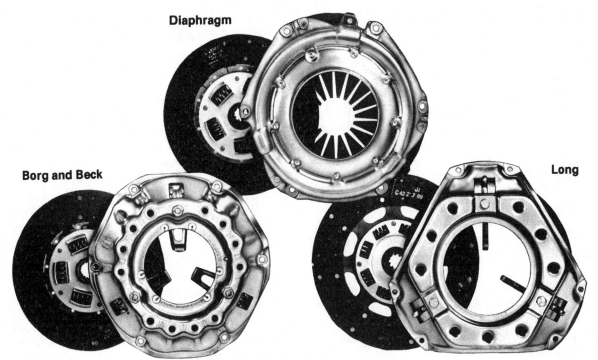

Diaphragm

Borg and Beck

Long

Illustrations courtesy of Borg Warner.

addition to the springs. Borg and Beck clutches generally have 12-spring pressure plates. The clutch cover distortion during release is a characteristic drawback of the Borg and Beck style. It is, however, the most popular competition style.

Long — The holding pressure in the Long style is a combination of static spring force and force from counterweighted release levers. The Long style always features centrifugal assist holding pressure. It generally has a 9-spring pressure plate. The long style features the advantage of high holding pressure under high RPM's but yet when disengaged, the lower static spring pressure means less force applied against the crankshaft thrust bearing. The long style also generally features a lower leverage ratio, which translates into quicker action for the clutch.

Diaphragm — This style features a circular, cone-shaped diaphragm spring. This spring contributes entirely to the holding pressure of this style of clutch. At high RPM disengagement, the release fingers of the diaphragm spring can be pushed so far forward that the centrifugal force can keep them against the flywheel. This makes the clutch stick in disengagement. While this is a characteristic drawback for diaphragm clutches, the overcentering and sticking can be prevented by using the proper thickness clutch disc and proper air gap between the disc and flywheel, and a stop on the pedal to limit travel. The clutch manufacturer will supply this specification, and you must keep a check on the clutch adjustment to help it live. Another drawback of the diaphragm clutch style is that it features no centrifugal force assists for holding pressure, so the spring pressure must necessarily be much higher. This has a very negative effect on crankshaft thrust bearing life.

CLUTCH DISCS

There are three methods of attaching the clutch facing material to the backing plate — bonding, riveting, and a combination of both. The bonding of the facing material to the backing plate is best for all-out racing. Rivets through the lining concentrate torque shear forces on the points immediately surrounding the rivets.

Performance clutches today are available with three differing kinds of lining materials — organic, semi-metallic and fully metallic. For better holding pressure and better heat dissipation for racing, at least a semi-metallic lining is recommended. Many Grand National racing teams are now using a fully metallic lining clutch.

Clutch disc marcel is a spring material which backs up the clutch facing material. It is obtained by installing a crushable space of spring steel sheet in a manner that looks much like the center section of corrugated cardboard. Marcel has its place in street clutches where preventing clutch chatter is more important than instantaneous lockup. But in competition clutches marcel is undesirable because it causes a lag in disengagement.

Very closely related to the marcel is the spring-cushioned center hub. In the street-use clutch, the sprung hub is very important to avoid driveline shock and unpleasant feedback shocks into the passenger compartment. In competition cars for years, the sprung hub was considered essential to the longevity of the U-joints and other components affected by quick lock-up shock loadings. But improvements in durability have pushed that reasoning aside, and now more and more solid hub clutch discs are showing up on oval track race cars. For short track racing where shifting is not important, we do not see any reason to recommend between the two types of clutch disc hubs.

DESIRABLE CHARACTERISTICS

With all of these facts behind you, it still is somewhat difficult to make a knowledgeable selection of the proper clutch for your application. To help you with your selection, let us add these desirable characteristics: First, you want the lightest rotating mass available. This allows the engine to be accelerated much quicker, and can be easier on the brakes when you back off the throttle. This desirable weight advantage applies to the flywheel as well as the clutch. Secondly, you want to have the greatest amount of lining area available, without the need of having too great a clutch diameter. The larger diameter contributes to a greater spinning inertia. Next, you want as much holding pressure as possible, yet with a fairly low static spring pressure (this combination means greater holding at speed, with less force applied to the crankshaft thrust bearing upon disengagement). Last, you do not want to sacrifice unit weight at the expense of pressure plate cover strength.

CLUTCH AIR GAP

The best and most effective method of adjusting the clutch is to measure the air gap existing between the clutch disc and pressure plate, or betwen the clutch disc and flywheel, while the clutch pedal is fully depressed against its stop.

The correct clutch adjustment is important to the proper operation of the clutch in high performance applications. If it is set too wide, you are wasting a lot of travel, which abuses the clutch linkage, and it also increases the time required to shift.

Before checking the air gap, adjust the proper clutch pedal height and the stop height. Check the throwout bearing-to-fork clearance for 1/8-inch to 1/4-inch. Lubricate all the moving links in the clutch release linkage, and be sure the linkage does not have any slop in it form worn parts.

To check the air gap, have a friend fully depress the clutch pedal. With a feeler gauge, check the distance between the flywheel and clutch disc, and compare that with the specification stated by the clutch manufacturer. once the proper gap has been obtained, disengage the clutch several times, then recheck the air gap.

Because the clutch disc is going to wear and decrease in thickness, recheck the air gap frequently. The linkage must be readjusted to maintain the proper clearance.

To simplify clutch adjustment, replace the rod between the clutch pedal and the clutch bellcrank with a steel tubular piece which is threaded to accept a spherical rod end bearing on each end. Then, the clutch can be adjusted from the driver's compartment by loosening a jam nut and lengthening or shortening the rod.

CLUTCH LINKAGE

In a racing application, a stock passenger car clutch linkage will probably require strengthening or replacing to accommodate a high pressure clutch assembly. To insure the proper disengagemennt of the clutch for high RPM shifts, no shaft or arm flexing can be allowed. Clutches which employ centrifugal force assists will especially cause problems with linkages which are not reinforced.

In constructing the linkage, you must consider linkage bind, pedal effort, linkage ratio and total travel of the throwout arm. When designing the linkage, remember that the engine is going to be moving slightly from side to side in the chassis because of torque reactions (rubber-mounted engines will move more than solid-mounted engines). The standard rod-type clutch linkage is partially anchored to the engine and partially to the frame. For this reason, the side of the block is fitted with a ball stud from which the cross shaft may rotate — even when the engine is torques over — without binding. The ball stud and socket fitment into the cross shaft must be maintained. The cross shaft may be altered — shortened, offset, etc. — but the flexible link between the frame and the engine must be maintained.

HYDRAULIC CLUTCH LINKAGE

The hydraulic clutch is the easy way out of having to put up with the problem-ridden mechanical linkage hardware. Engine movement does not affect the proper of operation of hydraulic clutch. The clutch actuating slave clyinder may be placed on the left or the right side of the engine, whichever is more convenient. The hydraulic clutch slave cylinder can be detached from the engine with one or two bolts, making engine changing much easier. Be sure the slave cylinder is positioned with its bore perpendicular to the fork so no side loads are placed on the piston seal.

When using the hydraulic clutch, be sure the master cylinder is securely mounted to the firewall so the firewall does not flex when the pedal is depressed. If it does flex, the clutch will not release properly. Also, be sure the slave cylinder is mounted to the block or bellhousing with a secure, properly designed bracket.

Because the clutch actuation will require a lot of fluid travel, use at least a 3/16-inch I.D. fluid line between the master cylinder and slave cylinder to keep the clutch operating at a normal speed.

A Chevy truck detached clutch master cylinder and slave cylinder is a good choice to use in a race car.

CLUTCH LINKAGE CHECKLIST

The clutch fork pivot should be held firmly in place. The clutch fork should be checked to see that it does not interfere with the bellhousing all through the travel. The clutch linkage bellcrank should be checked to see that no deflection or binding of it occurs when disengaging the clutch. Be sure no part of the clutch linkage interferes with

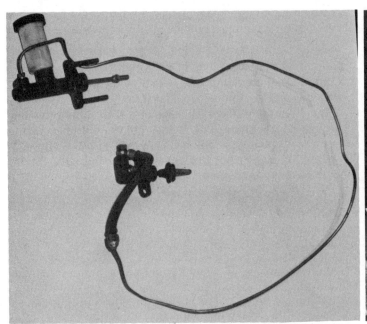

This is hydraulic clutch system from a Chevrolet truck which has been applied directly to a race car application.

Note the return spring on the slave cylinder push rod. This is a must if you want to prevent abnormal throw-out bearing wear.

headers, frame, etc.

CLUTCH INSTALLATION CHECKLIST

Start with an inspection of the flywheel face. It should be smooth, flat and clean. It is highly recommended that any time you install a new disc and pressure plate, you should have the flywheel resurfaced.

Check the pilot bearing or bushing in the end of the crankshaft for wear. Because this is such a small investment part and yet such an important prt, it is highly recommended that a new bushing be installed.

Next, be sure the clutch spline size is correct for the application by trying it on the shaft **before** installation. Grease the pilot bushing, the clutch shaft splines, the transmission extension sleeve and the yoke fingers at the point of contact with the bearing carrier.

When using the aligning tool to attach the clutch, be sure the lineup shaft is pushed all the way into the pilot bushing. Attach the pressure plate and torque it down in a pattern, with progressive torquing. Be sure the clutch disc is installed with the sprung hub facing away from the flywheel. Be sure your hands are clean when installing the clutch disc as any amount of oil or grease will detract from its proper performance. Use Loctite on all cap screws which attach the clutch to the flywheel.

TROUBLE SHOOTING CLUTCH FAILURES

Clutch malfunctions or premature failures are usually blamed on the clutch itself, and a new one (usually of a different brand) is installed without ever checking for the real cause of the trouble or failure.

One clutch manufacturer says that 85 percent of the returned prematurely failed competition clutches he sees are damaged due to throw-out bearing problems. Either the wrong part is used, the clearance is adjusted too tight, or an old or worn bearing is installed with a new clutch. In a race car, the throw-out bearing can wear very quickly. When a driver waits in line to qualify, etc., with the engine idling, he should shift the transmission to neutral and keep his foot off the clutch pedal completely. Even small pressure on the pedal will tke up the clearance between the bearing and arm, and keep the bearing spinning.

A worn clutch shaft will quickly ruin a new clutch hub, and lead to failure of the entire unit, because of excessive clearance between the splines which will cause shock loadings to the clutch.

Heat checked or scored flywheels or pressure plates will cause rapid clutch facing wear. The heat checks will act just like a file to wear away facing material. Always have a flywheel resurfaced before installing a new clutch.

When installing a new clutch disc, take the pressure plate out and lay a steel straight edge across the face of it. Check to be sure that the pressure plate has not worn convex or concave. If the pressure plate is warped, all of its contact with the disc will be at the inner or outer edge. This will result in slippage and rapid wear at the edge, and reduced holding pressure. Do not install a new pressure plate with a used clutch disc because the disc will have a tendency to wear the plate in a convex shape.

Misalignment between the flywheel and transmission, or the bellhousing and transmission, can cause many serious problems. The evidence is usually a broken clutch disc, unevenly worn clutch hub, failure of the pilot bushing, and grabbing and chattering of the clutch (because one side is

After the flywheel gets some wear on it, it should be checked for grooves, cracks or abnormal wear which can effect the clutch.

contacting first, not the entire face).

When installing a Long or Borg and Beck style clutch, lay the package on a flat surface and check the finger heighth. The manufacturer should provide you with the correct finger heighth. What it is important to check for, though, is that all three fingers are exactly the same height. If they are not, an adjustment is provided at each eyebolt nut.

MULTIPLE DISC CLUTCHES

The triple plate Borg and Beck brand clutch is the standard of the industry in multiple disc clutches. It is small in diameter, very lightweight, has a low rotating mass, has several plates for a large surface area, has a great holding pressure without a heavy spring pressure, and is very expensive. It is this type of clutch which enables Formula cars to shift so fast.

The smaller diameter means you can lower the engine in the chassis. The lighter weight and lower rotating mass means quicker engine response. And the use of more than one disc means an extremely long life for the clutch. In fact, the clutch should be good for at least two full seasons of racing, which might help in your decision to purchase one.

There are two companies which use the English-made Borg and Beck clutch as the basis for a complete package ready to bolt on to your Chevy small block or big block, or Ford small block. They include an automatic transmission starter ring so the stock starter can be used in the stock

location, and a special throwout bearing and clutch fork ball stud so the stock bellhousing can be used. The makers of these kits are Hoosier Racing Tires in Lakeville, Ind., and Quarter Master Industries in Elk Grove Village, Ill. Hayes Clutches — a division of Mr. Gasket — is also now making their own version of the triple plate clutch.

Borg Warner also makes a multiple disc clutch, this one being a double disc model. It does not quite measure up to the Borg and Beck brand, however. It is larger in diameter greater in weight, and has less friction lining area. It does cost less than Borg and Beck packages, however, so it might be considered as an alternative.

The Borg and Beck brand of triple disc clutch bolted up to a Chevy small block engine.

Race Prepping A Transmission

BUYING A USED TRANSMISSION

The first thing you will want to do is determine which transmission you have found, and the gearing in it. Our transmission case identification chart (courtesy of Hurst) in the Wrecking Yard Parts section will help you determine which transmission it is by brand and model.

To determine the gear ratio in each gear, shift the transmission in each gear and turn the input shaft one full turn, then count the output shaft revolutions. The revolutions is the gear reduction or ratio. Use a pair of vise grip pliers on the input shaft so it is easy for you to turn the shaft and count one full revolution. On the output, make a pencil mark on the housing and shaft to make it easy to count the revolutions. The accompanying chart shows all popular transmissions and the ratios they have been equipped with to aid you in your identification.

After you have determined what you have, you want to look at the mechanical soundness of the transmission. Start with the case. Look for cracks or a wet oiliness on the case. Try to avoid any transmission with a suspected case problem.

Transmission cases made for the Chevrolet bolt pattern (either Borg Warner or Muncie) have a tendency to have a cracking problem at the right lower corner. The aluminum or steel cases are worse than the nodular iron cases in this area. The problem stems from an anemic reinforcement of the bolt retaining ear at that corner. One solution to preventing this problem on Chevy cases is to drill the bolt hole one size oversize, install the bolt and draw it up snug but not tight, then safety wire the bolt to keep it in place.

The next thing you can do to determine the amount of wear and abuse a particular transmission has seen is look at the tail shaft bushing. Knock the rear seal out (you're going to need a new one any way if you are going to race the transmission). Look for the original thickness and chamfer of the end of the bushing. A bad bushing will have been knocked flat and have a sharp edge. Look for the original oil grooves and original finish of the bushing. A bad one will discolor and have marks in it. Also look for gouges and burnt spots. If the bushing is worn out completely and the slip yoke has marred the housing, a new bushing won't stay in. You will need a new tail housing, so skip that transmission and go searching for a more useable one.

WHICH TRANSMISSION TO USE

The four most popular four speeds in use in race cars are the General Motors Muncie, the Borg Warner Super T-10, the Ford T&C Toploader and the Chrysler New Process, in that order. The popularity takes into account such matters as availability and price, and fit, as well as the mechanical soundness and reliability of the particular transmission. In the following, we have analyzed the four transmissions for their desirability in terms of advantages and disadvantages.

Muncie — The biggest problem is that General Motors has discontinued the production of this very popular four

Be sure to check the brass synchro rings for cracks at each side of the key slot. Look for cracks in the strut keys too.

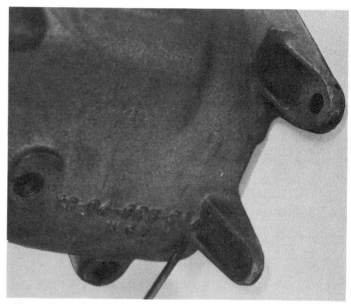

Chevrolet bolt pattern cases have a cracking problem at the lower right corner. Look for an oiliness or wetness in the area being pointed out to indicate a case crack.

speed, so the only source for it is from wrecking yards. But don't worry — Muncie made millions of them through the years. Mechanically, the biggest problem with the Muncie is the first gear freezing to the mainshaft. One popular fix for this problem is the removal of all the teeth off the first gear, and simply by-passing it. Another fix is putting a bronze bushing inside the first gear so it will not operate steel-to-steel. Problem is, though, if you get the bushing out of center in the hole by as little as .0002-inch (no, we didn't put in too many zeroes!), you can ruin the transmission in a hurry. There are a couple of transmission specialists who can do this work — Pepe Estrada at Phalanx Corp. in Paramount, Calif., and Tex Powell at Tex Enterprises in Asheboro, N.C. The best answer, though, is just be sure you have a minimum of .012-inch gear end play so the lubricant can circulate, and you shouldn't have any problems.

The Muncie is blessed with very big and heavy gears, which make it quite durable. But these heavy gears create quite an inertia problem when they are rotating, thus they are hard on the brass synchro rings. The rings tend to stretch after a period of competition useage in the Muncie. The Muncies built before 1971 have a 1-1/8-inch O.D., 10 spline input shaft. From 1971 on they ave a 1-1/8-inch O.D., 26 spline input.

Borg Warner Super T-10 — As a point of history we will point out that Borg Warner began making its original Warner Gear four speed back in 1957 as the T-10. It was a very widely used transmission. In 1970, Borg Warner introduced the Super T-10 which replaced the T-10, and had modifications in it which made it more competitive with the Muncie and Ford Toploader. The Super T-10 incorporates thicker and stronger synchro rings, a sharper tooth angle on the synchro teeth, and sturdier gears. The Super T-10 has a

weight problem however. In 1974, Borg Warner introduced what they called the Super T-10 Second Design, which incorporated many improvements over the super T-10. But, the Second Design Super T-10 was made for General Motors applications only, while B-W left the original design Super T-10 in their line for application in Ford and AMC cars.

The Second Design Super T-10 includes these improvements over the original design: 1) Nodular iron main case for 50% strength improvement, 2) Aluminum extension housing to shave ten pounds, 3) ¼-inch thicker mainshaft, 4) Greater strength gears, 5) Cluster countershaft diameter increased 1/8-inch, 6) No oil ring groove in cam selectors, 7) First gear is sleeved, 8) B-W's special high performance synchro hubs with no grooves come stock.

The GM application Second Design Super T-10 can be adapted to a Ford or Chrysler engine, with a little work, so the only drawback of this transmission remains the higher

This is a new tail-shaft bushing. When looking at a used transmission, the bushing in it should have the same chamfer and thickness on the end, and have the original oil grooves.

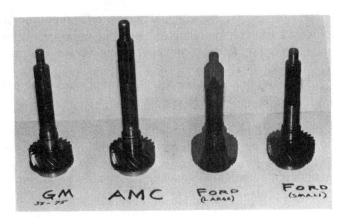

A comparison of input shaft sizes.

weight. Either of the Super T-10 models can be used for racing, but forget about the T-10.

Ford T&C Toploader — The T&C name comes from Thompson and Collins, the designers of this transmission. The Toploader designation comes from the fact that the transmission is easily identified by its 10-bolt top inspection cover. Strength-wise, the Ford Toploader has the biggest and widest gears available, and overall is a very durable and rugged transmission. It is desirable enough in racing that Toploader-to-Chevrolet bellhousings and pilot bushings are made.

The only area in a Toploader which could be worrisome to you is the mainshaft. Through 1974, they were made of a low alloy steel, then surface hardened. In competition, this particular mainshaft gave a lot of problems. Since early in 1975, however, Ford has been importing a high strength mainshaft from Mexico for the Toploader. If you want to find a high nickel alloy mainshaft replacement for your toploader, Borg Warner makes one.

The Toploader comes with two different sizes of input shaft: 1-3/8-inch O.D. 10 spline, and 1-1/16-inch O.D. 10 spline.

Chrysler New Process — The Chrysler four speed is a durable and reliable transmission, with a variety of gear ratios available. Its biggest drawback is that it is too difficult to adapt this transmission to a General Motors or Ford engine.

TRANSMISSION INTERCHANGE

There are three major factors effecting transmission interchange from one manufacturer line to another: 1) Transmission to bellhousing bolt pattern, 2) Input shaft spline and size, and 3) Overall length of transmission.

The third item can be eliminated when we are talking about transmissions for race cars. The overall length or the output shaft spline is not going to matter because a custom driveshaft for the car is going to be built anyway.

The bellhousing bolt pattern and input shaft length problem can be eliminated by using adaptor bellhousings

designed to accommodate transmission interchanges.

CHEVROLET SWAPS

Any Chevrolet engine from 1955 to 1971, from a Vega to a Corvette, with any engine, used the same input shaft spline, diameter and length. In the 1971 model year, a change was made making the input shaft a fine 26-spline version, and it also incorporated a larger diameter 32-spline output shaft (the same one as used in the Turbo 400 automatic transmissions).

The Borg-Warner improved version Super T-10 (which took the place of the Muncie M-22) will bolt up to the fine-spline application. The first design Super T-10 will bolt up to the course-spline applications in place of the Muncie transmission.

A Ford toploader can be installed behind a Chevy engine with little trouble. The bellhousing bolt pattern must be redrilled to accommodate the Ford transmission, or a special adaptor bellhousing can be purchased from Mr. Gasket Co. (part number 15060). A different pilot bushing is also required (part number 15975 from Mr. Gasket). The other modification required to bolt the Ford transmission up to the Chevy engine is to cut 3/8-inch off the length of the snout of the transmission (or clutch release collar). The spline pattern of the Ford unit is the same as the Chevrolet, but the Ford input shaft is 1/16-inch in diameter less than the Chevy. The Ford input shaft will fit with a Chevy clutch, but the fit is slightly sloppy, and the smaller diameter splines means the shaft is receiving high loadings up on the center of the spline ears. This swap has been successfully done before and run in competition, but we do not recommend it because of the extra loading on the input shaft. The best solution is to purchase a Chevrolet application clutch cover, and a Ford application clutch disc which will fit the pressure plate.

The Ford three speed Toploader has the same input shaft as the four speed, so everything explained here applies to mounting it to the Chevy engine.

FORD SWAPS

Putting a Muncie M-22 or Borg Warner improved version Super T-10 behind a Ford engine is basically the reverse process of what we just described above. The bolt pattern on the bellhousing must be redrilled to accommodate the GM pattern. The Ford pilot bushing can be resized through machining to fit. The input shaft spline will be too large in diameter to fit a Ford clutch hub, so a Chevy application clutch disc will have to be used in a Ford clutch cover.

BORG WARNER SWAPS

If you are going to use a Super T-10, find one from the same application as you will be using (for example, find a Super T-10 from a Ford application if you plan to use one with a Ford engine). It will save you a lot of work. If you order a new Super T-10, you will be specifying either Chevrolet, Ford or AMC application of it.

AMC SWAPS

The AMC transmissions — which are all supplied by Borg Warner — all use the early Ford (T-10) bolt pattern. The input shaft diameter and spline is the same as the GM course spline input shaft, but the AMC input shaft is longer and it requires a larger pilot bushing. Putting an AMC-application four speed behind a Ford or Chevrolet engine is more work than you need.

CHRYSLER SWAPS

Mr. Gasket makes a bellhousing adaptor and pilot bushing which allows you to mount a Chrysler New Process four speed behind a Ford or Chevrolet engine.

SYNCHRO RINGS

The purpose of the brass synchro ring in a transmission is to brake and slow the rotating mass of the input shaft (with its inertia of the spinning gears and clutch) to match the rotating speed of the next gear being shifted to. When you stop and think about the tremendous forces these parts encounter, you can see that they take a lot of abuse and are subject to a great deal of wear even under the most ideal conditions.

When you get your transmission opened up on your workbench, the first thing you want to check is the synchro rings. They should move freely radially up to the distance of the key slot. There should be a gap between the brass ring and the steel synchronizer teeth. If there isn't, the brass synchro ring is worn out. Also check the brass synchro rings for cracks at each side of the key slot.

The steel synchro teeth should be pointed. If they are not, it means somebody probably kept grinding the transmission into gear when the brass rings were bad. Means you have more parts to replace, too.

Look at the strut keys. Usually the ends or sides will break off if they are bad. Look carefully for a crack which can later develop into a broken key.

If you find a broken gear in a transmission, do not re-use ANY gear from it. Assume that a piece of metal has been squeezed between two gears and created stress cracks. All gears should be metal X-rayed, not Magnafluxed, when looking for flaws.

TRANSMISSION CHECKING AND REBUILD

There are three areas in the transmission which are critical to its proper operation as a racing component, and should be checked or modified to suit the requirements. These are the cluster gear end clearance, the gear end play, and the synchro sleeves.

CLUSTER GEAR END CLEARANCE

The gear end clearances and cluster gear end clearances are two of the most important things to check in a racing transmission if you want longevity and reliability from it. The clearances directly influence the lubrication circulation in the transmission. If you have only the minimal clearances, you will have only minimal lube circulation, which will cause heat build-up, which in turn will cause all sorts of other problems.

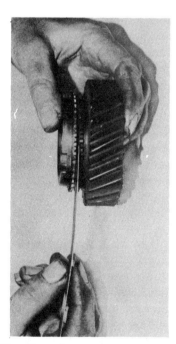

Checking the gap between the brass synchro ring and the steel synchro teeth.

Checking for gear end clearance. End clearance is controlled by snap ring thickness. If you use a feeler gauge like this, plan to have one set aside just for this job because you have to bend the feeler.

Left, the Ford Warner T-10 thrust washer [.108-inch thick], right, the Super T-10 thrust washer [.062-inch thick].

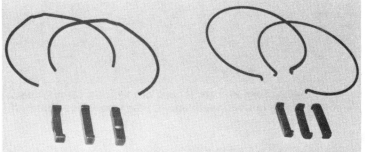

Different size snap rings to adjust gear end clearance are available in the Borg Warner Small Parts Kit. Use Borg Warner's solid strut keys, bottom left, not the hollow ones, bottom right. Below, checking for gear end clearance.

Checking the cluster gear end clearance in the case can be done in two different ways. Above right, a dial indicator is mounted on the case and bears against a gear to check the clearance. This is more time consuming, but more accurate. Below right, a feeler gauge can be used to check the cluster gear end clearance. With practice and care, this can be just as effective as the dial indicator method.

The cluster gear end clearance **must not be less than** .021-inch in an oval track or road racing transmission. It can tolerate up to .020-inch. Adhering to these numbers is vitally important. A drag racing transmission can tolerate much less clearance, and in fact it is desired. Drag racing transmission cluster gear end clearances are usually .005 to .010-inch.

The accompanying photos show how to measure for cluster gear end clearance. Be sure to do it. If you find the end clearance must be altered, change the thickness of the thrust washers. All thrust washers made for a particular transmission — for example the Super T-10 — are the "same" thickness as produced by the factory. But pick through a number of washers, mike them, and you will find a variation in thickness. The normal thickness is .062-inch for a Super T-10. However, you will find the washers as supplied by the factory to range from .060 to .065-inch.

If the end clearance is too small, have a machine shop surface grind the little end of the cluster gear whatever amount it takes to get .012 to .015-inch end clearance. Amazingly enough, the micro finish the surface grinder puts on the gear is very compatible with the thrust washer.

If you have too much end clearance in a transmission — for example a Super T-10 — you can get a thicker thrust washer from another transmission. For example, the Ford Warner T-10 has a .108-inch thick thrust washer, so it can be installed, and then have the small end gear surface ground to fit the desired clearance.

Whenever you have the transmission out of the car, periodically check the thrust washer on the cluster gear. If the washer has grooves in it, the end clearance is too tight.

GEAR END CLEARANCE

The gear end clearances are determined by the snap ring thickness. Check the gear end play on each for a minimum of .012-inch. If it is less than the minimum, the gear can seize to the mainshaft. To increase the clearance, decrease the thickness of the snap ring. Snap rings are supplied in the Borg Warner "Small Parts Kits" which are available for most all transmissions. There are a variety of snap ring thicknesses supplied with the kits, but they are not color coded so you have to mike them to identify them. If you are using the Small Parts Kit to rebuild a transmission, be certain to mike the snap ring thicknesses before installing them in the transmission. You just might be building in end clearances which are undesirable.

JUMPING OUT OF GEAR

Having a transmission jump out of gear is one of the most common ailments experienced with a transmission in oval track and road racing. There are some simple remedies, though.

Borg Warner makes a special high performance synchro sleeve which can be used with any T-10, Super T-10 or Muncie transmission. The B-W part number is T10S-15. For oval track racing, use these torque lock synchro sleeves for third and fourth gear. For road racing, use them for all four gears.

The torque lock synchro sleeve features a reverse ground taper on the interior of each tooth so as more load is applied to it, it has a gerater tendency to stay in gear. With the straight edge cut on regular gears, the shifting wears a taper into the gear which promotes jumping out of gear under load. The reverse ground taper in the B-W part is exactly the opposite of the taper which normally wears.

For oval track racing, another item which will help the jumping out of gear problem is to use solid strut keys. The Borg Warner Super T-10 comes with lightweight hollow strut keys which helps promote quick shifts. But in oval track racing these hollow strut keys will promote jumping out of gear too.

The third problem associated with inability to stay in gear is transmission misalignment with the bellhousing. If it is misaligned, the transmission input shaft will revolve in an oval pattern, putting side forces in a jerky pattern on the parts inside the transmission. This will put the tranmission right out of gear. Refer to the bellhousing segment in this chapter for more information on proper alignment

The special Borg Warner torque lock synchro sleeve, above. Compare its tapered teeth against the stock sleeve below it.

procedures.

LUBRICATION

There are several excellent transmission lubricants available on the market. We can recommend the following lubricants which we know to be excellent for the job: 1) Lubrication Engineers' Almasol #601 Trans-Worm Gear Lubricant (SAE Grade 90), 2) Ford's C2AZ-19580-D Hypoid Gear Lubricant (SAE Grade 90-140), and 3) Valvoline's #829 SAE 85W-140 Gear Lubricant.

No oil supplements need to be used with these recommended gear lubricants.

PILOT BUSHING

A pilot bushing is as good as a pilot bearing. There is little to be gained by switching the application to a bearing. The bushing is not heavily loaded, but rather it is just there to stabilize the input shaft against a wobble. Be sure a periodic inspection of the bushing is made for wear. A worn bushing will contribute to the transmission slipping out of gear.

THE BELLHOUSING

To prevent damage and wear to the transmission, it is very important to check the bellhousing alignment to the block and to the input shaft to be sure it is not off-center.

To check for the runout, install an old transmission input shaft (cut off about eight inches long) with a plate or bolt attached on the end. Mount a dial indicator on the input shaft and set the indicator plunger against the inner surface of the bellhousing (see photos for more explanation). Turn the crankshaft from the front of the engine 360 degrees (remove the spark plugs to make it easier) and watch the indicator for a runout difference. Whichever direction reads

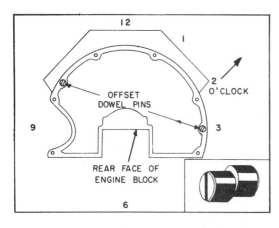

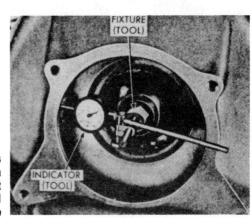

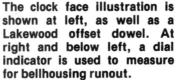

The clock face illustration is shown at left, as well as a Lakewood offset dowel. At right and below left, a dial indicator is used to measure for bellhousing runout.

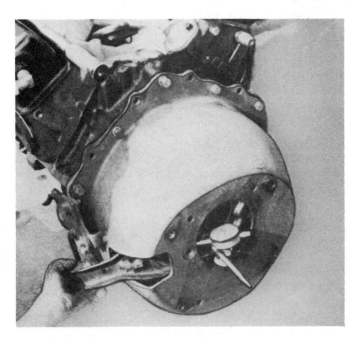

less in distance from the center of the input shaft, move the bellhousing in that direction by one half of the indicated error. For example, if you think of the bellhousing opening as being numbered like a clock face, and you find the bellhousing is off-center toward 3 o'clock by .014-inch, then move the bellhousing toward 9 o'clock by .007-inch.

There are two methods used to correct bellhousing runout. The first is to use offset dowel pins in the mounting. The problem is, though, a lot of times the offset dowels will not accommodate the problem. The easiest way to correct an alignment problem is to drill all the bolt holes and dowel holes 1/32-inch oversize. Then re-indicte and tap the bellhousing into place with a soft-headed hammer while the bolts are just snugged. Then tighten the bolts to the proper torque and re-indicate to be sure nothing moved.

The maximum tolerable runout on the bellhousing is .005-inch. Any more than this will create an input shaft wobble which will impair shifting and cause jumping out of gear.

If you have more than one engine block and bellhousing, you will find the oversize holes are the easiest way to correct alignment because the dowels have to be set differently when you mix the parts up. But, if you have just one bellhousing and one block, and you plan to take the bellhousing off only once or twice a season to check the clutch, then the offset dowels is the best way to go.

TRANSMISSION BREATHER

For Borg Warner tranmissions which have a plastic breather, knock it out and install a Ford rear end housing breather. For road racing, run a vent line from the transmission breather hole to a breather can mounted at dash board height to prevent lubrication slosh-out.

SHIFT LINKAGE

It is possible for a bad or weak shifter linkage to put you out of an oval track race. So, for extra insurance, install a heavy duty shifter and linkage.

If you are running a road course, it will pay great dividends for you to choose a shift linkage which is as stout as possible. This means using the largest diameter rods available, and steel shims and washers instead of nylon ones. We've seen some strong, over-anxious drivers actually mangle some weak shifter linkages on a road course. We recommend the Hurst Super Shifter with reverse lock-out for road courses. Hurst has worked with racers for a long time and has provided excellent technical services and many specialized parts such as steel bushings, etc.

Transmission Gear Ratios

Transmission	1st Gear	2nd Gear	3rd Gear	4th Gear
B-W Super T-10 2nd Design	2.64	1.61	1.23	1.00
B-W Super T-10 2nd Design	2.64	1.75	1.33	1.00
B-W Super T-10 2nd Design	2.88	1.74	1.33	1.00
B-W Super T-10 2nd Design	3.44	2.28	1.46	1.00
B-W Super T-10 2nd Design	2.23	1.77	1.35	1.00
B-W Super T-10 2nd Design	2.43	1.61	1.23	1.00
B-W Super T-10 2nd Design	2.43	1.76	1.47	1.00
B-W Super T-10 2nd Design	2.64	2.10	1.60	1.00
B-W Super T-10 2nd Design	2.64	2.10	1.46	1.00
Warner T-10	2.54	1.92	1.51	1.00
Warner T-10	2.10	1.64	1.31	1.00
Warner T-10	2.54	1.89	1.51	1.00
Warner T-10	2.36	1.78	1.41	1.00
Warner T-10	2.36	1.76	1.41	1.00
Warner T-10	2.73	2.04	1.50	1.00
Warner T-10	2.36	1.62	1.20	1.00
Chrysler New Process	3.09	1.92	1.40	1.00
Chrysler New Process	2.66	1.91	1.39	1.00
Chrysler New Process	2.65	1.93	1.39	1.00
Chrysler New Process	2.47	1.77	1.34	1.00
Ford Toploader	2.71	2.04	1.51	1.00
Ford Toploader	3.16	2.22	1.41	1.00
Ford Toploader (heavy duty)	2.32	1.69	1.29	1.00
Ford Toploader (heavy duty)	2.78	1.93	1.36	1.00
GM Muncie	2.20	1.64	1.28	1.00
GM Muncie	2.52	1.88	1.46	1.00
GM Saginaw	3.11	2.20	1.47	1.00
GM Saginaw	2.85	2.02	1.35	1.00
Warner T-16	2.41	1.57	1.00	—
Warner T-16	2.86	1.71	1.00	—
Chrysler A-230	2.55	1.48	1.00	—
Chrysler A-230	3.08	1.70	1.00	—
Ford Toploader	2.42	1.61	1.00	—
Ford Toploader	2.99	1.75	1.00	—
GM Saginaw	2.54	1.50	1.00	—
GM Saginaw	2.84	1.68	1.00	—

Transmission Input Shaft Data

Maker	Input Shaft
GM (1955 to 1970)	1-1/8'' O.D., 10 spline
GM (1971 & later)	1-1/8'' O.D., 26 spline
Ford (all)	1-1/16'' O.D., 10 spline
AMC (all)	1-1/8'' O.D., 10 spline
Chrysler Hemi, 440	1-3/16'' O.D., 18 spline
Chrysler 318, 340, 360, 383 (1961 and later)	1'' O.D., 23 spline

The Rear End

THE FLOATER REAR END

In the interest of safety, a full floating rear end should be used in any race car. Full floating means the axles transmit torque only, and they do not retain the hub or bear any of the vehicle's weight. If an axle should break in a full floater, the wheel will not come off as will happen in a semi-floating rear end as found in passenger cars. The full floating rear end will cost you some money to purchase for your car, but it is an investment that will pay you dividends many-fold in terms of durability and safety.

AXLES

When you purchase a full floating rear end or quick change floater, you will also be purchasing a set of axles to go with it. These axles, for the utmost in durability, should be made from 4340 steel, be heat treated for strength, shotpeened to relieve surface tension that could cause a crack, then micro finished to provide longer life.

If you have two axles of the same length in your rear end, never interchange the axles from side to side after they have been used. Axles will take a set in the direction of their normal rotation, and changing sides will create a stress reversal which could lead to an axle failure.

A completely locked rear end doubles the torsional stress on axles, so be sure you have heavy duty axles when using any kind of a locker.

THE FORD 9-INCH REAR END

The Ford 9-inch rear end is the standard of the industry for stock car racing, and so, we have concentrated heavily on it in this chapter. It is called the 9-inch rear end because of its use of a 9-inch diameter ring gear. This rear end is very popular for racing use because it is very strong and reliable, is easy to assemble, is quite simple in construction, and uses heavy duty parts. A variety of gear ratios is widely available for it as well. The removable carrier section makes gear changes and routine maintenance very easy.

When you are looking for a 9-inch rear end, look for one which has a nodular iron carrier housing. It can be identified by an "N" cast into it just above the pinion retainer. Look for these under some of the older Ford muscle cars, such as the 1968-70 Mustang and Cougar with 428 or 429 engines, or 1968 to 1970 Torinos and Montegos with 428 and/or 429 engines. If you cannot locate one, order one from Ford under part number D00Z-4141-A. The nodular iron is much higher strength than regular cast iron, and definitely should be used for a high performance application.

Many nodular iron carrier housings were also made without the characteristic "N" cast into it. There is another way to tell the housings apart, though. There is a casting number cast into the housing between the gasket surface and the carrier bearing saddle on one side. The correct casting number you should look for is C4AW-B.

The Ford housings also come equipped with 28-spline and 31-spline axles. The 28-spline carriers are more plentiful in the wrecking yards, but you will save yourself some time and money if you buy the 31-spline variety. It definitely is what you want to use to go racing with. A 28-spline rear end

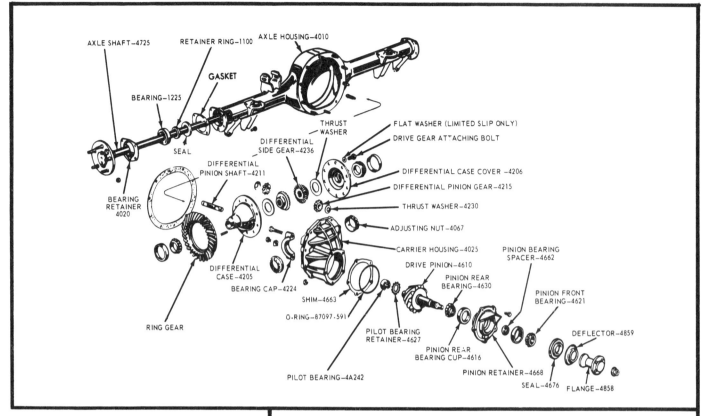

Above, a parts breakdown of a stock Ford 9-inch ring gear rear end. At right, the durable Ford rear end reworked as a full floater for racing.

can be converted to a 31-spline quite easily. See details in the Locked Rear Ends section of this chapter.

SETTING UP THE FORD REAR END

We will start this discussion assuming you have purchased a rear end — either new or used — which has the incorrect ratio ring and pinion in it. We will proceed through teardown, checking and reassembly.

Before you start, make yourself a check list to be sure you have these items on hand: a selection of pinion retainer shims, a selection of pinion bearing spacers, a new housing gasket, a new pinion seal, a foot-pound torque wrench, an inch-pound torque wrench, a magnetic base dial indicator, and a gear marking compound (such as white titanium dioxide marking compound available at auto parts stores).

Place the pumpkin on your workbench. (We will refer to the removable carrier which contains the ring and pinion as the pumpkin. The front section of it which carries the pinion we will call the pinion retainer. The back section of it which carries the ring gear we will call the carrier housing). Use a center punch and a hammer to make a mark on one of the carrier bearing caps and its corresponding bearing housing.

This matches the carrier bearing caps for you as sets. This is important during reassembly because the bores are align-bored and the caps must not be swapped from side to side. Remove the adjusting nut locks, bearing caps and adjusting nuts. Pull the ring gear and its case out. Make sure the carrier bearing races are marked to be sure they go back with the correct bearings.

The pinion retainer can then be removed from the carrier housing. Using a pipe wrench to hold the companion flange, the lock nut on the pinion can be loosened, then the pinion driven out of the retainer with a hammer. Use a chisel and hammer to drive the pinion seal out of the retainer case, then pull out the front pinion bearing and pinion bearing spacer. Wash the parts in solvent thoroughly (providing a check of a used bearing indicates it won't have to be replaced with a new one), then air dry the parts. Don't wipe them off with a rag because the lint particles will damage the bearings. Don't spin the bearings with air pressure when drying — you can exceed their designed speed.

Press the inner pinion bearing off the pinion, and press the bearing back on the new pinion. Be sure to wash the bearing in solvent and check it for wear.

While everything is apart (after everything has been washed in solvent), deburr and chamfer the ring gear teeth, deburr the carrier, deburr the saddle caps, and polish the gear teeth lightly with a 3M Scothbrite General Purpose Wheel (#5A Fine). Clean the ring gear bolt threads with a spray spark plug cleaner.

The pinion and pinion retainer is the first assembly to go back together. Before we proceed any further, we should point out that there are two different size pinion retainers and rear pinion bearings. Older housings were equipped with a thick-flanged retainer which used a larger O.D. inner bearing. It is no longer serviced for replacement parts by Ford. If you get hold of a rear end equipped with a large bearing carrier, you can switch to the small bearings by purchasing and installing the small carrier, bearings and adjuster nuts. The I.D. on all the bearings is the same.

To assemble, insert the pinion into the retainer, then place the pinion bearing spacer (we'll cover the correct size in the next paragraph) on the front side, followed by the bearing cup, bearing, and companion flange. Tighten the lock nut on the pinion to 175 foot-pounds. This will draw the bearings together and seat them. Check the pinion bearing preload by rotating the pinion with an inch-pounds torque wrench. Your ultimate goal is 5 to 8 inch-pounds without the oil seal. It shouldn't reach that point yet. If it has, disassemble and choose a thinner spacer. Your ultimate goal for lock nut torque on the pinion is 180 to 200 foot-pounds, but you don't want to tighten it up all at once without checking the bearing preload, or you will Brinnell the bearings. Be sure not to use an impact wrench for the same reason.

The pinion bearing spacer is the item which controls the pinion bearing preload. For a given housing and gearset, the thicker the spacer, the lesser the preload. The variable in determining the spacer thickness is the machining of the pinion retainer. Thus, a correct spacer thickness will always remain with the same pinion retainer, regardless of the gearset. A solid spacer can be reused indefinitely. Choose a pinion bearing spacer of the same thickness as came out of the housing when it was disassembled. If you do not know what thickness the spacer was, start with one .480-inch thick. They are available from Ford in 20 sizes ranging from .465 to .485-inch. The .480 is a fairly safe starting point as most assemblies end up using a spacer within .002-inch of it, so there is little chance of harming the bearings as the lock nut is torqued.

Install a new O-ring on the pinion retainer, then the pinion depth shim, and install the assembly onto the carrier housing. Lubricate the O-ring before installing it and be careful not to twist it.

The pinion depth shim used for assembly should be .019-inch thick. This is a very average median starting point, and usually the adjustment for pinion depth will require very little change one way or the other. Ford makes a variety of shims in various thicknesses, from .466 to .490, available under the basic part number of 4662.

Remove the old ring gear from its carrier, and check the carrier for any burrs which might keep the new ring gear from seating properly. Check the back of the ring gear for burrs also. Assemble the ring gear onto its carrier, using Loctite Stud-Lok on the threads of the bolts, and tighten the bolts in an alternating pattern. Drop everything into place, install the bearing caps in the correct positions, and just snug the cap bolts. Screw the adjusting nuts so they just contact the bearing cups, and be sure they are turned in the same number of threads on each side. Be sure the bearing cups are lubed so they move easily. The adjuster nuts will determine the backlash clearance, and the amount of carrier bearing preload. The backlash is the amount of clearance between the ring and pinion gear teeth. It is designed into the gears to permit the flow of lubricant on the face of the gears, and to allow the gears to grow in size slightly when operational temperatures rise.

The adjuster nuts have both been run in equally until the bearing cups have been touched. Back off the adjuster nut on the ring gear side three turns, then tighten the pinion side

At left, the arrow points to the casting number location which indicates the good nodular iron case. Right, a wide variety of ring and pinion gear ratios are available for the Ford 9-incher, which makes it a very attractive unit for racing.

After everything is assembled, a pattern check can be made. The ring gear has been smeared with titanium dioxide marking compound and the pinion rotated [with some pressure applied against it] to obtain the pattern. This pattern looks perfect.

so it shoves the carrier and ring gear away from the pinion. Then back off the adjuster nut on the pinion side three or four turns. Running the adjuster nuts in, then backing them off, is an effort to center the bearings in the races. When you back off the adjuster nuts, the races stay centered. It is very important to get the bearings and their races centered before you set and check the backlash.

Then screw the adjuster nut on the ring gear side in to push the ring gear towards the pinion. While you are carefully doing this, gently rock the ring gear back and force

to feel and listen for the backlash between the gears. Keep on screwing the adjuster nut in until the ring gear reaches the point of no backlash being felt (remember that the other adjuster nut on the pinion side must be backed off enough so there is nothing to stop the ring gear from moving its way). As soon as the backlash has disappeared, stop screwing on the adjuster nut on the ring gear side. Then screw the adjuster nut in on the pinion side and watch the bearing race on its side through the 5/16-inch hole on the top of the bearing cap. Tighten the adjuster nut until it contacts the bearing race. When it contacts that race, it will make the race start to rotate. The instant it starts to rotate, tighten the adjuster nut 2½ more holes. This sets the preload. Setting the bearing preload on the pinion side has pushed the ring gear slightly away from the pinion, giving you a good starting point for the backlash. Once you feel the carrier bearing preload is set, check it with an inch-pounds torque wrench. To do this check, you must remove the pinion shaft from the case. With Ford gears, it should be 8 to 10 inch-pounds. With Zoom or Schiefer gears, it should be 10 to 15 inch-pounds. If the preload is not correct, set it with the adjuster nut on the pinion side.

At this point, mount the dial indicator and check the backlash. The dial indicator base should be clamped to the gasket surface of the pumpkin. Point the indicator plunger against the edge of a tooth at exactly a 90-degree angle to the tooth surface. The indicator must be perpendicular to the ring gear radius or you will get an error when you read the backlash. The backlash you are looking for is: .008 to .009-inch with low gears (6.00 to 4.11 ratios), .010 to .012-inch with medium gears (4.11 to 3.50), and .013 to .014-inch with high gears (3.50 and less). Check the backlash at a minimum of three places around the ring gear.

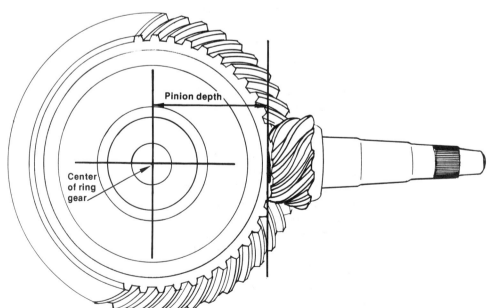

This is a comparison of the ring gear and pinion in relationship to the pinion depth.

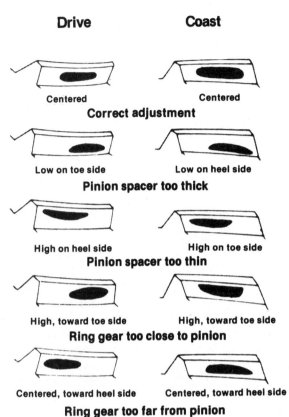

Drive   **Coast**

Centered   Centered
Correct adjustment

Low on toe side   Low on heel side
Pinion spacer too thick

High on heel side   High on toe side
Pinion spacer too thin

High, toward toe side   High, toward toe side
Ring gear too close to pinion

Centered, toward heel side   Centered, toward heel side
Ring gear too far from pinion

Because of light loading on the gears, it is sometimes difficult to interpret the pattern. Look for the heaviest contact part of the pattern for your interpretation.

A comparison of ring gear contact patterns. An interpretation of both the drive and coast sides of the gear tooth is included. Note that if the tooth contact pattern is very narrow, all across the length of the tooth and near the top of the tooth, the teeth will wear thin or break at the top.

There should be **no** variation. If there is, suspect a warpage somewhere that may cause some type of binding or unusual loading in the gear operation.

If you find that you need more backlash, start by backing up the adjuster nut one half hole on the ring gear side, and then tighten the pinion side adjuster nut one half hole to keep the pinion bearing preload uniform. If you need less backlash, reverse this process. Once the backlash and preload is set, torque the bearing cap bolts to specification.

Once the unit is assembled, you are ready for the gear pattern check to see the pinion depth into the ring gear. Remember that the depth is controlled by the thickness of the pinion retainer spacer. The pinion depth is the distance from the ring gear axis to the back face of the pinion gear. This measurement is important to establish the correct contact area between the ring and pinion teeth.

To get a depth pattern check, paint the ring gear with white or red titanium dioxide marking compound (white is best for visibility). As the pinion gear teeth contact the ring gear, a pattern will show up and tell you if the ring gear is running true and the pinion has been spaced right. Check the depth mesh first, then look for the contact pattern. To check the pattern, you need to turn the ring gear with a wrench on one of the ring gear attaching bolts, and apply pressure to the pinion area by wrapping a shop cloth around the smooth part of the yoke, and hold a certain amount of pressure on it. Turning the thing free won't show much of a pattern. Consult the accompanying chart and drawings to

interpret the patterns you find on your ring gear.

With the lower gear ratios typically used for short track racing, the pinion gear has less teeth and is smaller in diameter. This makes for a much higher unit loading on each gear tooth. This means that lower gear ratios are more prone to breakage, so they are much more critical for proper setup of the gear contact pattern. With a low gear ratio, don't settle for a pattern which is "just about right." It must be perfect, even if it takes a little more time.

Once the gear mesh pattern is satisfactory, the pumpkin is ready for installation into the axle housing. Install it using Ford part number B7A-4035-A housing gasket. It is an asbestos-composition gasket. Don't use any other brand of paper gasket or any sealant on the gasket. If you do, you will be in for leakage problems. Be sure to use Loctite on the two bolts that hold the tabs which hold the adjuster rings.

THE DIFFERENTIAL

When a car with a standard differential negotiates a corner, the outside rear tire covers a greater distance, thus it rotates slightly faster than the inside rear. This is a "differential" or difference in speed. The action of the outside rear tire is also called overspeed. In racing, the overspeed of the outside rear is not nearly as important as it is in passenger cars making sharp corners at low speeds. Take for instance a car with a wheel track of 60 inches (5 feet), cornering on a 200-foot radius turn (typical of a short track turn). The outside tire will require 5/200 or 2½ percent

FORD 9" RING GEAR SETS		
Mfr	Part #	Ratio
Ford	C4AZ-4209-A	3.10
Ford	C4AZ-4209-B	3.40
Zoom	180024	3.89
Zoom	180025 (8620 steel)	4.11
Zoom	189007 (9310 steel)	4.11
Zoom	189008 (9310 steel)	4.29
Ford	C4AZ-4209-C	4.33
Zoom	180026 (8620 steel)	4.57
Ford	C4AZ-4209-E	4.71
Zoom	180027	4.71
Ford	C4AZ-4209-F	4.86
Zoom	180028	4.86
Ford	C4AZ-4209-G	5.14
Zoom	180029	5.14
Schiefer	7290017	5.14
Ford	C4AZ-4209-H	5.43
Zoom	180030	5.43
Schiefer	7290005	5.43
Ford	C4AZ-4209-AD	5.43
Ford	C4AZ-4209-AE	5.67
Ford	C4AZ-4209-J	5.67
Zoom	180031	5.67
Schiefer	7290007	5.67
Zoom	180060 (8620 steel)	5.83
Zoom	189012 (9310 steel)	5.83
Schiefer	7290019	5.83
Moroso	8571	5.87
Zoom	180061 (8620 steel)	6.00
Zoom	189009 (9310 steel)	6.00
Moroso	8572	6.00
Schiefer	7290021	6.00
Zoom	180062 (8620 steel	6.20
Zoom	189011 (9310 steel)	6.20
Moroso	8573	6.20
Schiefer	7290023	6.20
Moroso	8575	6.50

The worst method of achieving a solidly locked rear end is by welding the spider gears. Welding on these heat treated high alloy steel gears can cause stress cracks in them which will break under operating conditions. And you can bet for sure that extraneous pieces of gear, or even weld slag, rotating about in the differential will chew things up in a hurry!

The best and cleanest technique of solidly locking the rear end is to use a machined spool to replace the spider gear assembly. The spool will also help reduce the weight and rotating mass. A spool can also be used to convert from one spline pattern axle to another. To convert a 28-spline Ford to a 31-spline, bore out the ring gear carrier ends 1/8-inch oversize, install a locking spool of the appropriate size, and bolt everything together. It is a good idea to replace the carbon steel pinion shaft with a chrome moly steel item.

On very short radius turns, such as a small quarter-mile paved track, hard acceleration will probably produce some amount of understeer. The problem can usually be balanced out of the chassis with spring rate changes, or a slight amount of wedge adjustment. If the problem cannot be cured with these changes, go back to the original chassis starting point, then use a slightly larger circumference tire on the outside rear corner (this is called tire stagger). Using tire stagger for the problem, put the larger diameter tire on the right rear.

The greatest disadvantage of the solidly locked rear end is when an axle breaks, or when one tire loses traction on oil, water, etc. The forward thrust remaining on the other tire will cause a tremendous leverage around the car's center of gravity, resulting in a spin. Even on a straightaway, it could cause a virtually impossible-to-control spin.

THE DETROIT LOCKER
In a Detroit Locker-equipped rear end, any time that overspeed or differentiating must take place, the outside rear wheel is disengaged and it rotates freely. Under

This is the mini-spool locker, made by Speedway Engineering. The most sensible way to solidly lock a rear end. Note the chrome moly cross pin.

greater speed than the inside tire. But the tire slip rate (caused by load transfer when entering the turn and by thrust when exiting) is greater than 2½ percent, so the overspeed effect is negligible.

LOCKED REAR ENDS
Because of the tire slip rate in a race car, it can operate with a solidly locked rear end. This has the advantage of delivering complete forward thrust from both rear tires, regardless of the load condition placed on each tire and the coefficient of friction each tire encounters. There are several ways of locking the rear end.

Left, a properly mounted rear end housing vent. Above, a run-out indicator [arrow] is used to check housing straightness of a floater rear end. Any rear end used in a race car must be checked for straightness so a variable is not introduced into the suspension geometry.

straightaway conditions, the Detroit Locker solidly locks both wheels together. The big advantage of the Detroit Locker for oval track or road racing is that it eliminates the problem of locker-caused understeer going into and coming out of a turn.

The only major service problem with a Detroit Locker is that the springs inside it lose their tension after a while. The symptoms of this problem are that the car will drive or veer to one side because one axle is not locking up completely. To check the problem, disassemble the Locker and check both the heighth and spring rate of the two springs on a valve spring checker against a new spring.

REAR END VENT

The rear end housing must be positively vented, or you will find lube sloshed all over. Run a ¼-inch I.D. oil resistant hose from the vents and route them high into the forward part of the trunk. Terminate the hose ends in a can or breather.

DETERMINING HOUSING WIDTH

Usually when you order a floater rear end housing from a manufacturer, all you have to do is give him the desired track width you want as well as the wheel width and wheel offset you will be using, and he will be able to compute the flange-to-flange length required.

QUICK CHANGE SPUR GEARS

To get a much longer life and reliability from quick change spur gears, grind down the ends of the teeth about 5 degrees on a belt sander. The surface hardening process these gears go through causes a stress build-up on the ends of the gears. All spur gear breakage eminates from the outside. After grinding, chamfer the edges of the gears to remove sharp edges and flash. The last process is to touch the edges of the gears with the 3M Scothbrite General Purpose Wheel (#5A Fine) to polish them.

LUBRICANT

If you have a differential which is 95 percent efficient (which is a very high level of gear reduction efficiency), it is absorbing five percent of the horsepower being transmitted from the drive line into it. The horsepower is absorbed by friction, and friction means heat, and the heat must be absorbed and carried away by the lubricant. This should tell you that any old oil is not sufficient to get the job done.

In addition to its heat absorbing requirement, the lubricant must also provide a solid film between parts to prevent metal-to-metal contact. In the hypoid type of gearset such as is used in most rear ends, the gear teeth slide together as the teeth contact, making a very effective press to push the lubricant off of the gear teeth. Hypoid gears require a lubricant designed for extreme pressure application to prevent gear pitting.

The following lubricants have the ability to meet all the requirements of a good rear end lube, and give good life and service to the parts: 1) Lubrication Engineers' #601 Trans-Worm Gear Lubricant, 2) Ford's C2AZ-19580-D Hypoid Gear Lubricant, and 3) Valvoline's #829 SAE 85W-140 Gear Lubricant, and 4) Phillips 66 gear lubricant, and 5) Union 76 High Performance Gear Lubricant.

REAR END COOLING

A rear end lubricant cooler is needed if the lube temperature consistently gets over 240 degrees F. This can be checked by touching a pyrometer probe to the housing lube area, or by attaching colored thermal strips to the housing.

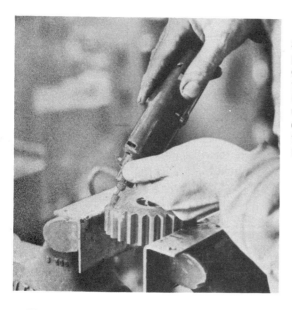

Sharp edges and flash are a problem on quick change spur gears, as seen above left. The flash is carefully ground off, left. At right, a belt sander is used to grind a 5-degree angle on the ends of the teeth. The finished product is seen above right.

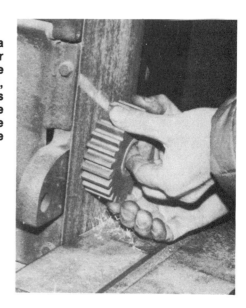

The rear end cooler, if used, should employ a Jabsco "Water Puppy" (part number 6360) to pump the lube. The lubricant should flow from the housing to the cooler to the pump and back to the housing. This flow route avoids overheating of the pump by the lubricant.

GEARING

The terminology of gearing in reference to a "high gear" or a "low gear" can get confusing. A high gear ratio is one which is numerically low, for example 2 to 1. The "high" means that the car will move a further distance for each engine revolution. A "low" gear ratio is numerically high — such as 6 to 1 — but means a slower speed but greater torque at the same engine RPM's.

There is a formula which can be used to estimate the final rear axle ratio which is required for a particular oval track or road course. It is:

$$\text{Axle ratio} = \frac{\text{(Engine RPM) (Static tire radius)}}{\text{(Car speed) (168)}}$$

In this formula, the engine RPM is an average of the RPM the car will be turning as it pulls off a turn and the maximum RPM desired at the end of the straightaway. An example of this would be a car which comes off a turn at 4500 RPM and the maximum desired RPM is 6800. The average would be 5650 RPMs. The static radius is the center of the wheel to the ground. The car speed (expressed in MPH) is the maximum MPH expected to be attained around the entire course. To make this easy, use the qualifying record (in MPH) for the track.

For an example of the use of the formula, let's say the average RPM is 5650, the static tire radius is 12.75 inches, and the qualifying record is 71.112 MPH. Then the desired axle ratio would be 6.03 to 1.

If you are close to being in the ballpark, but need to gain or subtract a few RPM's, here is a rough rule of thumb (for pavement tracks) which can help you out: Every 100 RPM's is .10 on a gear ratio. One inch of tire circumference is worth a change of 120 RPM's.

Welding Basics

We are not going to teach you how to weld in this chapter. Welding is much more extensive a subject than can be covered just briefly here. It has been covered thoroughly in several books, which you may want to read. They are available from a well-stocked welding supply store, or your public library. They can supply you with the basic knowledge. You can learn the techniques and the intricacies of the subject at a trade school or junior college, or have a certified welder teach you.

ARC WELDING

We will deal more extensively with arc welding here than oxy-acetylene welding, because one or more of the forms of arc welding will be most extensively used in building your race car.

The basic arc welding process is the electrode stick arc welder. An arc is created when an electric current, regulated by a welding transformer, flows across an air gap between an electrode and the base work metal to be welded. The intense heat generated by this arc is suited for welding as it can be directed to affect only the part of the work metal to be welded. To be assured of uniform heat from the arc, the arc length must be kept the same for a given rod size and current setting.

The instant the arc is struck, a part of the base metal directly below it is melted, resulting in a pool of molten metal. Some of this molten metal is blasted out by the arc and is deposited along the path of the weld. The depth of the crater formed by this blasting action is the distance the weld will extend into the base metal, and is referred to as the penetration of the weld.

The electrode is also melted simultaneously with the base metal and is carried by the arc to the molten pool. This added electrode metal combines with the base metal to form the deposited weld. During the transfer of metal from the electrode, the flux coating on it burns off and forms a gaseous screen that completely envelopes the arc, protecting the molten metal from the harmful effects of oxygen and nitrogen in the surrounding atmosphere. The remainder of the melted flux removes impurities from the molten pool and deposits them on the surface of the weld as slag.

The TIG (tungsten inert gas) arc welding method was the next to be developed. It uses a non-consumable tungsten electrode, and was originally developed to production-weld very thin and delicate materials in the aircraft industry. Its arc is protected from impurities in the atmosphere by a flow of argon or helium gas around the arc. Argon and helium are inert gases, which means they will not mix with any other substances, so they provide a solid shield for the arc. TIG welding is also used with a filler material which is fed into the molten puddle from the side of the torch by hand, much as in oxy-acetylene welding. TIG (or heliarc) welding is used in race car construction where 4130 steel tubing or aluminum sheet sections are used in the construction. This is most notably in drag cars and Indianapolis cars. The welds are very high quality, but are very time-consuming to make.

MIG welding (which stands for metal inert gas) is basically a refinement of the TIG system. It uses a continuous-feed filler rod material pulled through the torch from a roll of wire. The filler material is also the electrode. The protection from the atmosphere in MIG welding is provided by a shield of carbon dioxide or argon gases, or a combination of both. The MIG — or wire welder as it is sometimes called — is the most versatile type of welder for building a race car, and it is the easiest type of welding to learn, and perform.

ADJUSTMENTS TO THE WELDING PROCESS

The weld can be regulated or adjusted for the type of work material and thickness of it by regulations to the current. Before we examine how these regulations affect the weld, we should define the terms:

Alternating current (or AC) reverses or alternates its flow direction a specified number of times in a second. The number of times it reverses is called the **frequency**. The most common form of AC current is the one which makes a complete change of flow from one direction to the other in 1/60 of a second. The complete change is called a **cycle**, so there are 60 cycles per second. A more modern term replacing cycles per second is **hertz**. Thus, the most common form of AC current is 60 hertz. There are also 25, 30, 40 and 50 hertz AC currents in existence.

Direct current (or DC) flows continuously in one direction, never reversing itself. Direct current can be reversed, however, either with a switch in the welding machine, or by switching the poles which attach to the work material. **Reverse polarity** is achieved by connecting the electrode cable to the positive pole on the machine, and the cable from the work piece to the negative pole on the machine. **Straight polarity** is achieved by connecting the electrode cable to the negative side of the machine, and the work piece cable to the positive pole on the machine.

Amperes (or amperage) measures the intensity of the current flow in the circuit (or, the rate of flow). Welding machines are usually rated by the maximum amperage it can supply.

Volts (or voltage) is the pressure or force which pushes the electrical current through the circuit. **Arc voltage** is the amount of voltage actually used in the welding process.

Voltage drop — Just like water being pushed through a pipe, the farther water or electricity travels in a circuit, the weaker the pressure becomes. The weakness of the electricity is called the voltage drop. Also effecting the drop is the resistance the current encounters in the circuit. This is called the **ohms** of resistance.

Duty cycle — is an important consideration in the specifications of a welding machine which you intend to purchase. It is defined as the percentage of a ten minute period during which the machine can continuously operate safely at its rated power output. For example, if you have a 200 amp machine and it has a 50 percent duty cycle, you can continuously weld for 5 minutes out of every 10 at 200

amps. If you are using only half the amperage, however, you can safely operate the machine continuously. In race car fabrication, it is extremely rare for any machine to be operated even five minutes continuously. Thus, you can purchase a welder with a lower duty cycle without worries. This is important, because you pay a premium for a high duty cycle.

USING THE ELECTRICITY

The amperage creates the heat in an arc. The voltage creates the force which keeps the arc operating. Considerable voltage is required to keeep the arc operating.

There are definite advantages and disadvantages to both AC and DC current. A standard table available from a welding supplier will help you judge which type of current to use for the type of metal and thickness of material you will be welding.

Generally, AC current is considered to be a high production welding current. You can use thicker electrodes and greater amperage with AC. This results in a faster travel speed. On the other hand, AC is not suitable for welding thinner gauge metals. The thinner gauge requires a lower amperage, and AC is not as stable at the lower amperage.

DC, on the other hand, allows greater penetration of a weld. DC has a disadvantage of creating arc blow, however. Arc blow is the tendency of the arc to deflect from the path it is intended to follow because of magnetic field build-ups. It is characterized by excessive spatter and poor fusion.

All of these problems associated with current choice refer to the stick arc process. If you use the MIG process, you will be using direct current reverse polarity exclusively. This current produces a deeper penetration and a cleaner weld.

WELDING WITH MIG

As we have touched on already, the MIG welder has all the attributes of a welder required to assemble and fabricate a race car. While it is not as precise as heliarc (TIG), it is faster. The minimum thickness of material that can be welded with a MIG welder is 20 gauge (.028-inch thick), although an experienced operator can successfully weld thinner sections. For most sheet applications, straight carbon dioxide can be used as the inert gas, and this is a saving because carbon doxide is the least expensive of the inert gases. The MIG welding torch can be used in virtually all positions, and there is no worry of spatter if welding overhead. The penetration of the bead can be varied by the heat setting of the machine (amperage setting) and the speed at which the operator moves the torch. Different filler wires can be purchased to enable the welder to tackle most all materials, although a .030-inch diameter nickel alloy-based wire will get the job done for almost any application in a race car, from light sheetmetal to 2-inch thick plate.

MILD STEEL ARC WELDING ELECTRODES

Electrode Classification E4510	Coating sulcoated	Position*	Current**
E4510	sulcoated	F, V, OH, H, HF, F	DCSP
E4520	sulcoated	F, V, OH, H, HF, F	DCSP
E6010	High cellulose sodium	F, V, OH, H	DCRP
E6011	High cellulose potassium	F, V, OH, H	AC, DCRP
E6012	High titania sodium	F, V, OH, H	AC, DCSP
E6013	High titania potassium	F, V, OH, H	AC, DCSP
E6015	Low hydrogen sodium	F, V, OH, H	DCRP
E6016	Low hydrogen potassium	F, V, OH, H	AC, DCRP
E6020	High iron oxide	HF, F	DCSP, AC
E6030	High iron oxide	F	DCSP, DCRP, AC

* Position abbreviations
F — flat
V — vertical
OH — overhead
H — horizontal
HF — horizontal fillets

**Current abbreviations
DCSP — Direct current straight polarity (electrode negative)
DCRP — Direct current reverse polarity (electrode positive)
AC — Alternating current

Arc welding electrodes are classified for their specific use by a four digit number. The number always starts with "E" which means electric welding. The first two numbers in the four digit number indicate the tensile strength of the electrode metal. For example, 45 means a minimum of 45,000 psi tensile strength, 60 is 60,000 psi, etc. The third digit is the welding position for the electrode. The fourth digit indicates the type of coating on the electrode as well as the welding current to be used with it. If the last digit is zero, then the third designates the coating and the current. An example of a typical electrode would be E-6015. The 60 indicates 60,000 psi minimum tensile strength, which would be for mild steel material. The 1 indicates all welding positions. The 5 indicates a low hydrogen sodium coating, to be used only with DC reversed polarity.

Arc Welding Rod Thickness	
Rod Diameter	Amperes Required
1/4	200-300
3/16	140-200
5/32	110-150
1/8	70-120
3/32	40-70
5/64	20-40
1/16	20-40

Beating Economy Stock Rules

RULE	SOLUTION
1. No special racing spindles. Passenger car spindles only.	1. Use the Cadillac spindles discussed in the spindle section.
2. No ballast may be bolted to the frame or chassis.	2. Solidly anchor the ballast inside the frame rail at the left rear corner and then plate over the frame channel. Or, install the ballast inside the roll cage tubes which attach to the left rear corner before they are welded in place.
3. The frame must be stock, and not lowered.	3. Cut wedges in the frame rails and reweld it, as shown in the chassis fabrication section.
4. The alternator must be wired and operable.	4. Install a simple on/off toggle switch in the field pole wire to turn the alternator off. In this way it offers no drag.
5. Pump gasoline only allowed.	5. Use the alternative gasoline outlined in the fuel system chapter. It will check at the same specific gravity as pump gasoline.
6. Steering must remain stock.	6. No problem. The stock linkage will be sufficiently strong. Use a 16 to 1 ratio gear to quicken the steering (see spindle and steering chapter for details).
7. Only 60 or 70 series tires permitted.	7. If 60 series tires are allowed, by all means use them. Use nylon bias ply construction tires. Use a tire durometer to find the softest tire compound. Shave the tread to one half the new tread depth.
8. No special racing anti-roll bars allowed.	8. Choose one from the list of production anti-roll bars listed in the chassis chapter.
9. No weight jackers allowed.	9. Use equal rate front springs cut to 10 inches tall, and equal rear coil springs cut to 12 inches. If weight bias is required to be jacked to one corner, or chassis height needs to be changed at one corner, use various sized spacers to set on top of each spring. If the spacers are disallowed, use springs of the same rate at varying heights — for example, have available an 850 #/'' spring at 8, 9, 10 and 11 inches tall.
10. Production passenger car only upper A-arms are allowed.	10. Production A-arms present no problems in mounting or fabrication over the tubular A-arms. The only draw-back is just a little extra weight. Cut the A-arms at a place where they are relatively straight in order to shorten them.

Testing A Race Car

The race car is finished! It has been built the best you know how, according to all the theories you have read about. Now it is time to see how it performs.

This brings us to a word of warning. Most people build a race car with a goal of finishing it in time to enter a specific race (without a specific goal, most of us would never finish a race car, right?). Most of the time this goal is met by finishing the race car about two hours before the first race begins. Everyone hopes it runs according to plan, that everything is put together right, and the guess on the chassis setting is about halfway in the ballpark!

That's not the proper way to go racing. Nothing should be left to chance. We can recite a lot of stories about race cars that were brought directly to the track after being finished in the afternoon, with disastrous results. Like the car that hit the first turn wall head-on after the throttle stuck wide open on the first hard lap. Or the car that backed into the wall on the first lap because the brake linkage locked up the rear brakes only. Or the car that caught on fire on the backstretch because a fuel line was loose. We could go on.

The point is, all of these disastrous occurances happened because the car was hastily assembled, and not tested. In a testing session, these problems could have been caught early in slower check-out laps, and corrected.

The race car should not be tested at an oval race track but rather at a drag strip and skid pad (you will usually find the two types of courses at one facility). The drag strip can be used for acceleration and braking stability tests. These tests will quickly show if the car is in sound mechanical condition, if anything is going to come loose, if the throttle is going to stick, and if the car is going to stop, and stop in a straight line. All of these tests are best conducted at a drag strip where there is a safety margin for shutdown or spinout

Proper testing of a race car can help avoid all the heartaches of such problems as a stuck throttle.

The idea of skid pad testing is to hold the vehicle steady and constant at the edge of control in a constant circle. The test procedure is helpful with any type of vehicle, running on any type of track, because it helps establish a baseline for the chassis settings.

should something go wrong (like the stuck throttle or rear brakes locking up).

The skid pad is nothing more than level, smooth area of asphalt paving where a car can be held in a constant circle (with a minimum radius of 100 feet). The skid pad tests for steady state cornering, and can very quickly sort out problems of oversteer and understeer, front end alignment, tire pressures and weight distribution. The skid pad must be marked off with a painted line to give the driver a precise path to follow. If you are making a skid pad out of the drag strip's or local shopping center's parking lot, the circle can be painted with a washable poster paint that can be washed away after the testing session.

TESTING PHILOSOPHY

The first step in testing is finding a base of reference. That means you start your test by recording every possible variable which can be changed and which can affect the performance of the race car. Then the first test is run with those specifications to give a feedback on how the car performs. All subsequent tests are performed as a comparison to the base test, to determine if one specific variable change makes a positive or negative change in the performance of the car.

In order to evaluate the value of any change, only one change at a time can be made. If a number of changes are made at one time, and the result is a faster lap time, you will not know why the lap time was faster. It may even be that two of the changes you made may slow the car down and only the third one will produce more speed.

The driver should be objective and honest. If a change is made which he thinks should make the car better, but in fact it does not, he should accept that fact and report it as such. He should also be sensitive to every action of his car so he can accurately communicate the feel of the car and its problems to his crew. At the skid pad is also a time when the driver can watch the action of the oil pressure and fuel pressure gauges during cornering, and he can report any

low readings or fluctuations to his crew.

On a skid pad, the driver must be consistant in his lap times. This is not a practice session. He should be driving the same line the same way on every lap, concentrating to stay right on the painted circle all the way around it at the very edge of tire adhesion. The only changes in lap times should be produced by changes in chassis settings.

Timing of the skid pad laps should be done with a reliable stop watch which shows time to at least a hundredth of a second. Using two mechanical stopwatches (stopping one and starting the other simultaneously) can produce a small error in getting the thumbs synchronized. Because skid pad testing is carried out on such a small circle, every hundredth of a second can be quite meaningful, so every attempt to eliminate error is important. Better than mechanical stopwatches is an electronic timer with a memory, such as the Siliconix Accusplit III watch. It also saves time because the read-out is digital. This gives the record keeper time to record a lap time and record a comment about each lap. These comments, such as if the car bobbled, or the driver went inside or outside of the painted line, can help in evaluation of the data collected. The timed readings for each lap must be taken as the car passes a certain point, such as a line painted on the track.

Electronic stopwatches such as these are the most useful instruments for timing a vehicle on a skid pad.

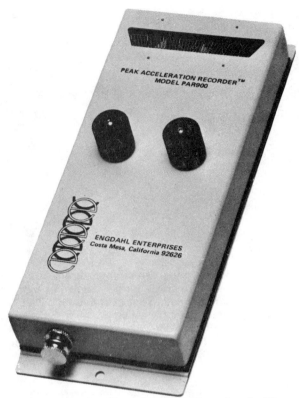

If you do not have access to a skid pad testing facility, you can use this mechanical accelerometer available from Steve Smith Autosports to mount in your car to measure the maximum G's of lateral acceleration attained.

In evaluating the times collected after a test run is completed, throw out the highest and the lowest times (plus any unusually slow times due to driver or car error), then average the rest of the lap times. This will then be the assigned lap time value for that particular test. Before taking timed laps for the car at the beginning of a test, give the car at least five good laps to get the tires warm. Only then should test data times be recorded. The lap times can be converted to G's of lateral acceleration by using the formula: G = 1.22 x radius/(time)² where the radius is expressed in feet and the time in seconds.

AN ACTUAL TEST

To describe the events contained in a skid pad test, we will present an actual test we performed on a 1969 Chevelle late model sportsman car. The methods of testing and evaluating data collected can be copied and used by anybody, with the result being a very competitive car the first time it hits the race track.

STARTING SPECIFICATIONS
1. Total weight: 3380
2. Weight per wheel: RF-900, RF-730, LF-950, LR-800
3. Weight distribution: 54.6% front, 51.6% left
4. Sprung weight: RF-781, RR-542, LF-831, LR-602
5. Unsprung weight: RF-119, RR-198, LF-119, LR-198

6. Roll centers: Front-½ '', Rear-11¼ ''
7. Motion ratios (squared): Front-.326, Rear-.438, Anti-roll bar-.916
8. Spring rates: RF-750, RR-300, LF-900, LR-250, Anti-roll bar-469
9. Wheel rates: RF-245, RR-131, LF-261, LR-110, Anti-roll bar-430
10. Front roll couple: 79.5%
11. CGH: 15½ ''
12. Front end alignment: Caster — RF + 3¾ , LF + ½
    Camber — RF -4¼ , LF + 1¼
13. Chassis height: 5 inches at each corner (measured from frame rails)
14. Shock absorbers: (Monroe part numbers) RF-87, RR-87, LF-86, LR-87
15. Tires: all Firestone 43's, static radius 13.5 inches
16. Tire pressures: RF-25, RR-23, LF-17, LR-21
17. Tread width: Front-66'', Rear-65½ ''
18. Wheelbase: 112''
19. Final gear ratio: 7.83 with locked spool
20. Starting track temperature: 71 degrees

If you do not understand the meaning or significance of some of the specifications, or changes, encountered in the test procedure, reading this book will help you gain an understanding.

Before the car was run, it was brought up to operating temperatures and all fluid levels were checked. The car was thoroughly checked for any leaks, then slowly driven to check for clutch, brake and accelerator pedal operation.

FIRST TEST
Lap time 11.20 seconds. This is the baseline test. Driver felt car was fairly neutral steering. Tire temperatures showed very cool on left front, and that inside of right front was too warm. Change: right front camber changed to negative 3¼ degrees.

		LF		RF	
95	95	95	210	152	150
		LR		RR	
118	115	115	145	150	150

TEMPERATURE CHECK

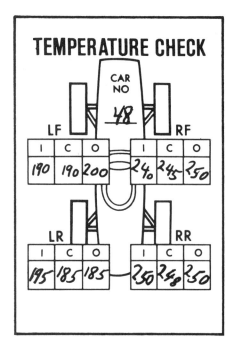

CAR NO 48

LF				RF		
I	C	O		I	C	O
190	190	200		240	245	250

LR				RR		
I	C	O		I	C	O
195	185	185		250	248	250

The numbers underneath each test indicate the tire temperature readings after the test was run. A regular tire temperature check form is shown at upper left. Above, this is the basic type of race car we used in our skid pad test. The facility we use is the Mira Loma Testing Grounds in Mira Loma, Calif. Below left, in "1" is shown the way to mount the accelerometer shown on page 163 to measure cornering G's. 2 shows how to mount it to measure braking and accelerating forces.

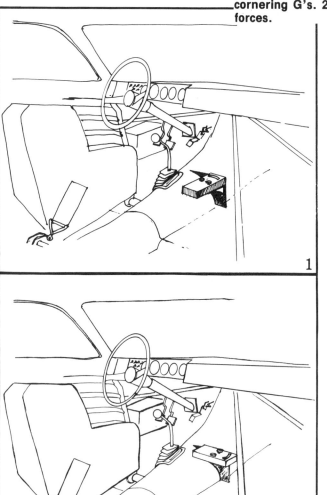

SECOND TEST

Lap time 11.15 seconds. The left front tire heat is still disappointing. The front tire temperatures seem to indicate that some toe-out is needed. Change: toe-out changed to ¼ " out.

	LF			RF	
96	92	90	173	140	150
	LR			RR	
108	107	105	118	118	112

THIRD TEST

Lap time 11.11 seconds. The toe-out change seemed to even out some of the heat on the front tires, but the right front still shows signs of having too much negative camber. Driver says car is pushing and getting worse with each successive lap. Change: right front camber changed to negative 2¾ degrees. Camber change automatically brought toe-out to 1/8" out.

	LF			RF	
92	95	92	135	122	118
	LR			RR	
92	101	105	118	118	115

FOURTH TEST

Lap time 11.04 seconds. The lap times indicate the camber change helped, the driver says the car understeered only a little, and the tire temperatures look better. We are happy with the front end alignment, so now we want to get more heat in the left front tire (which means it will be working harder). Change: 1150 # /'' spring installed at left front.

	LF			RF	
101	101	100	178	175	175
	LR			RR	
98	100	104	125	125	125

FIFTH TEST

Lap time 11.14 seconds. The car slowed down by lap time indication. The left front generated more heat, but at the same time both right side tires generated more heat. The driver says the car just wanted to skate sideways — it really didn't want to bite with either of the right side tires. Change: 700 # /'' spring installed at the left front.

	LF			RF	
112	112	110	182	175	172
	LR			RR	
105	115	115	148	146	148

SIXTH TEST

Lap time 11.20 seconds. Times are slower, the driver says the car understeered. Tire temperatures show right front tire needs a little more air pressure and the right rear needs a little less air pressure. Change: installed original 800 # /'' spring at the left front. This brings us back to baseline. To try to take the understeer out of the chassis, a 350 # /'' spring is installed at the right rear.

	LF			RF	
112	112	109	160	150	160
	LR			RR	
105	115	115	140	146	141

SEVENTH TEST

Lap time 11.18 seconds. Rear spring change brought the right rear tire temperature up, which according to the textbook would make you believe the car was oversteering. Driver says, though, that chassis felt quite unstable — it teetered between oversteer and understeer every lap. The variance between the lap times indicate that the car didn't feel right to the driver and he was fighting it. The thinking of the crew was that the car was fairly balanced with the spring rates at each wheel that the car was equipped with on the first test, but the front anti-roll bar was too stiff. So, the right rear spring was changed back to the baseline of 300 # /''. Change: 364 # /'' bar installed at the front.

	LF			RF	
96	96	95	152	156	155
	LR			RR	
117	118	119	147	148	147

EIGHTH TEST

Lap time 10.80 seconds. The lap time and tire temperatures seem to indicate that a positive change was made. Driver was happy with the feel of the car. The car was thought to be well balanced for over and understeer. The thoughts then changed to weight distribution. The car is nose heavy, and does not have enough static weight bias to the left. Using the weight jackers, the weight distribution was changed to 1000 at the LF, 860 at the RF, 770 at the LR and 750 at the RR.

	LF			RF	
114	114	112	178	176	176
	LR			RR	
119	120	120	159	159	158

NINTH TEST

Lap time 10.68 seconds. The car is very responsive to just a small weight change. The tire temperatures show that the greater left side bias gets those tires working more, despite the car being just slightly more nose heavy.

	LF			RF	
121	122	122	173	172	173
	LR			RR	
118	119	119	156	157	157

TENTH TEST

The next series of tests were moved to a 180-degree paved, flat turn which simulates the turn one and two of a 3/8-mile oval. Long straightaways enter and exit the 180-degree turn. Timing was done from a spot in the center of the turn across from the entry of the turn to the exit of the turn. The purpose of these tests is to determine if stiffer or softer shock absorbers help the chassis. This test is a baseline with the shock absorbers mentioned in the specifications. Lap time 5.29 seconds. Change: An 85 Monroe shock installed at the left front in place of the 86.

ELEVENTH TEST

Lap time 5.53 seconds. The car was very loose from the middle of the turn, all the way out of the turn. The 85 shock puts too much weight on the right rear tire as throttle is applied. Change: One Monroe number 86 shock absorber installed at the LF, LR and RR corners in place of existing shock absorbers.

TWELVETH TEST

Lap time 4.74 seconds. The softer shocks allowed better tire compliance and a better bite. Lap times were faster, the driver said the car felt very stable and driveable. End of testing session.

Race Car Check List

We have provided a checklist here as an aid to the racer in caring for his car. The important point is that proper preparation, organization and attention to details are most times the key to winning races — not the factor of brute horsepower or driver skill. It has been said before, and we will say it again — a car has to finish a race to win it.

We all can recount stories we know of about drivers who have lost races because they have failed to check oil levels, brake fluid levels, wheel bearing conditions, critical hardware tightness, etc. It all happens through carelessness and/or forgetfuness. Our check list is intended as one step towards eliminating those problems.

The list secondly is a check which should be followed through before the car goes racing each week. Sure, it can be said that some of the items are elementary and may insult the intelligence. But we can answer with the stories of well-known professional racing teams which have sent their cars into competition without fueling the car, or having any lube in the differential. A rigidly adhered to check list probably would have let those cars finish a race.

You might want to make up lists for yourself which you can tape to your windshield each week to be sure everything is checked.

WEEKLY MAINTENANCE

A. Cockpit
1. Are gauges operating properly?
2. Belt and harness condition and mounting
3. Seat mounting
4. Fire extinguisher — charge condition and mount hardware
5. Interior wiring
6. Cracks developing in roll cage tubing or welds
7. Tachometer cable tight
8. Fuel pressure gauge line tight
9. Cockpit clean, including glass or Lexan
10. Mirror mount solid

B. ENGINE
1. Oil level
2. Oil condition
3. Water level
4. Sparkplugs — gap and wear
5. Plug wires
6. Distributor cap — inside pole wear, cracking, arc tracks
7. Ignition high tension leads
8. Points
9. Valve lash
10. Pulley belts
11. Crank dampner
12. All external fuel and oil lines

13. Water hoses
14. Valve covers
15. Radiator core
16. Oil cooler — core and connections
17. Headers
18. Fuel filter
19. Fuel line from cell
20. Oil filter
21. Air filter
22. Throttle linkage
23. Throttle return springs
24. All carburetor bolts and screws — tightness. Also check tightness of boosters.
25. Carburetor fuel pressure
26. Cylinder head bolt torque
27. Compression check
28. Leak check
29. Timing
30. Fan bolts tight
31. Fan belt tension

C. SUSPENSION
1. Brake shoes or disc pads wear
2. Brake drum condition or rotor condition
3. Disc brakes
   a. Caliper bolts tight
   b. Hat bolts tight
4. Wheel cylinder condition
5. Wheel bearings condition, lube and preload
6. Brake system bleed
7. Master cylinder fluid level
8. Shock absorbers condition and mounting condition
9. Trailing arm bushings and hardware
10. Panhard bar — hardware and rod end condition
11. Panhard bar brackets — cracking
12. Crossmember cracking
13. Spring seats
14. Wheel studs condition
15. Upper and lower A-arm bushings condition and lube
16. Ball joints condition
17. Springs — true rate
18. Anti-roll bar
    a. Mounting bushings condition
    b. Spherical bearing links condition
19. Upper and lower A-arms attaching bolts
20. Steering linkage
21. Tie rod ends condition
22. Wheels condition
    a. Cracks
    b. Valve stems condition

D. ELECTRICAL

1. Battery charge condition
2. Battery connections
3. Battery securing hardware

E. DRIVE LINE
1. Clutch condition
2. Clutch free play
3. Clutch linkage condition
4. Clutch slave cylinder leakage
5. Clutch master cylinder fluid level
6. U-joints condition
7. Transmission leakage
8. Transmission case cranks
9. Transmission crossmember condition
10. Rear end leakage
11. Shifter linkage
12. Throw out bearing
13. Rear end lube level and condition
14. Transmission lube level and condition

F. CHASSIS GENERAL
1. Ballast secure
2. Bumper attachments secure
3. Fuel cell secure
4. Fuel line condition
5. Exhaust pipe brackets condition
6. Brake lines condition
7. Sheetmetal condition
8. Appearance
9. Hood pins and hood pin cables secure

G. CHASSIS ALIGNMENT
1. Weight at each corner
2. Chassis height at each corner
3. Caster
4. Camber
5. Toe-out

H. AT THE TRACK
1. Check lug nut tightness — most important
2. Spark plugs
3. Timing
4. Tire pressures
5. Fuel level
6. Oil level
7. Tire temperatures after warm-up

STEVE SMITH AUTOSPORTS
PUBLICATIONS
The Best in How-To Performance Books

The Stock Car Racing Chassis

The basic stock car handling "get acquainted" book. Covers: Understeer and oversteer • The ideal chassis • Spring rates and principles • Wedge • Camber/caster/toe • Sorting a chassis with tire temperatures • Building a Chevelle chassis. **#S101 . . . $6.75**

Complete Stock Car Chassis Guide

A more advanced stock car handling book. Covers: Lateral acceleration and weight transfer • Finding the CGH • Roll centers • Roll couple • Using anti-roll bars • Racing tire basics and selection • Shock absorbers selection • Aerodynamics basics • 2-point rear suspension. **#S102 . . . $6.75**

Bldg A Race Car Picture By Picture

A collection of detailed photos taken underneath, inside and under the hood of some of America's most successful stock cars. Shows the elements that make a truly winning car: Electrical and plumbing detail, construction details, component placement and much more. **#S110 . . . $6.95**

Racer's Complete Reference Guide

The "Yellow Pages" of the high performance world. Where to get: Hardware, chassis and engine parts, running gear, etc. Where to find: Fabricators, engine builders, safety gear, racing associations, driving schools, etc. A unique and valuable reference source for racers. 430 illus. **#S108 . . . $7.95**

How To Run A Successful Racing Business

How to operate a racing-oriented business successfully. How to: Start a business • Put more profit in a business • Cost controls • Proper advertising • Dealing with employees, taxes, and accountants • And much more. The secrets of business fundamentals and how to make them work for you. For existing businesses too. **#S143 . . . $9.95**

So You Want To Go Racing?

Finally, a book written for the enthusiast considering his first start in racing. Gives a complete understanding of: Setting up a shop • Buying the right tools and equipment • Organizing and gathering a crew • Getting the most out of a limited budget • Being resourceful • Shop setup • Budget checklists • And more. This book is like taking an apprenticeship to a 20-year racing veteran. **#S142 . . . $9.95**

Stock Car Driving Techniques

Written with authority by GN driver **Benny Parsons.** Covers: Basic competition techniques • Getting started in racing • Improving your lap times • Defensive driving tactics • Every aspect of competition driving. **#S104 . . . $7.95**

Advanced Race Car Suspension

The latest tech information about race car chassis design, set-up and development. Weight transfer, suspension and steering geometry, calculating spring and shock rates, vehicle balance, chassis rigidity, and much more. **#S105...$8.95   Work Book for above book . . . #WB5 . . . $5.95**

Building The Hobby Stock Car

For the low-buck racer with limited facilities. How to build a car with a minimum of "store bought" parts or expensive machine work. Covers: Car choice • Cage construction • Unitized vs. full frame chassis • Engine, cooling and electrics • Transmissions and rear ends • Suspension (parts choice, building and sorting) • Driving • Preparation, and more. **#S126 . . . $8.95**

Building A Street Stock Step-By-Step

Step-by-step info from buying the car and roll cage kit to mounting the cage, stripping the car, prep tips, front and rear suspension modifications, adjusting the handling within street stock rules, at-the-track setup, and mounting seats, gauges, linkages, etc. A complete guide for the entry level street stock class. **#S144 . . . $9.95**

Hot To Hotrod & Race Your Datsun

The famous Datsun high performance guide by Datsun specialist Bob Waar. For the 240, 260 and 280Z, 510, 610, 710 and 200SX. Contains: Complete Datsun engine building and blueprinting tips • Low buck performance tips for street machines • High performance engine prepping • Complete handling and suspension for track and street • Bolt-on power High performance parts list. And much more. **#S141 ... $15.95**

The Complete Karting Guide

The complete guide to kart racing, for every class from beginner to enduro. Contains: Buying a kart and equipment • Setting up a kart — tires, weight distribution, tire readings, plug reading, exhaust tuning, gearing, aerodynamics • Engine care • Competition tips • And more. **#S140 . . . $9.95**

Race Car Graphics

Takes the "magic" out of race car coloring, numbering, painting, striping and graphic design by showing how you can do it. Make your car and operation look absolutely professional without spending a lot of time and money. Includes 8 pages of full-color photos. **#S118 . . . $7.95**

Buick Free Spirit Power Manual

Complete information on the performance engine of the future — the Buick V6. Blocks, crankshafts, rods, valve gear, intake and exhaust systems, ignition, suspension, brakes and body modifications. Includes chapter on building on IMSA Skyhawk. Complete list of performance hardware. Over 200 photos and drawings. **#S123 . . . $9.95**

The Racer's Tax Guide

How to save BIG money on your racing activities. This book tells you how to LEGALLY subtract your racing costs from your income tax bill by running your operation like a business and following 3 simple steps. An alternative form of funding your racing. Includes a concise, up-to-the-minute report on how the new tax laws and changes help you save even MORE money. Read this book **now** and start saving! It works! **#S116... $9.95**

Race Car Fabrication & Preparation

Includes thorough discussions of: Chassis and suspension fabrication and design • Preparing a trans • Setting up a rear end • Electrical system • Building the roll cage • Cutting costs and beating the economy stock rules • Welding • Clutches • Wrecking yard parts • Plumbing • Driveshafts • Wheels and tires. Plus much, much more. **#S114 . . . $10.95**

Racing Engine Preparation

By **Waddell Wilson,** one of the best engine builders in NASCAR. He takes you through all the steps of engine building as carefully as he builds one. Explains: Cam timing and selection • Head porting • Carb mods • Prepping rods, pistons and blocks • H.P. ignition and lube systems. "The most complete engine book I've ever seen," says **A.J. Foyt.** Uses small block Chevy, Ford and MoPar. **#S106 . . . $9.95**

Racing The Small Block Chevy

A totally up-to-date high performance guide to the small block Chevy. Covers: High performance at a reasonable cost • Component blueprinting • Revealing cam, carb head porting and ignition tips from many of racing's best known engine builders • 330 illustrations. **#S112 . . . $9.95**

Bldg The Chevy Sprint Car Engine

Includes complete info on the Donovan and Milodon aluminum blocks, all aluminum heads, fuel injection systems and set-up, burning alcohol, magnetos, plus complete chapters on blueprinting all components. **#S130 . . . $8.95**

Bob Bondurant High Perf Driving

A fascinating book that teaches the necessary skills for high-speed driving at minimum risk. Bondurant translates track-proven techniques into 12 chapters about getting the most out of yourself and your car and controlling it! Learn to do it from the master. Over 100 illus. 144 pages. **#MB35 . . . $11.95**

Sprint Car Technology

A complete book on buying, building and racing a sprint car, midget or super modified. Covers: Torsion bars • Chassis structure • Straight front axle suspension • Live rear axles • DeDion rear • Birdcages • Lateral locating linkages • Steering • Bump steer • Shocks • Dirt and asphalt set-ups • Tire selection • And more. Photo packed. **#S125 . . . $11.95**

Dirt Track Chassis Technology

Covers EVERY facet of dirt track racing and stock car chassis set-up. Includes all the latest hot topics — left side bias, aerodynamics, leaf vs. coil suspension, and details on the 5th coil-over suspension. This massive new book is a MUST! **#S133 . . . $12.95**

Stock Car Chassis Technology

A brand-new suspension book by **Steve Smith!** The latest in race car suspension tips, all the newest ideas in stock car chassis set-up. Includes: Chassis set-up principles • Procedure for set-up at the shop, at the track • Actual case studies • Front suspension set-up and adjustment • Rear suspension — all the new torque absorbing systems, torque arms, stagger, locating devices • Repairing a crashed car • The newest on aerodynamics for the short tracks • And much, much more. For paved and dirt track applications. **#S139 . . . $12.95**

Stalking The Motorsports Sponsor

The methods in this clear, precise book will help you set up a sponsorship program whether you race on a local, regional or national scale. **#S119 . . . $7.95**

Chevy Heavy Duty Parts List

It is up-to-date with all the newest stuff from the factory. Complete with all engine and driveline goodies, plus much more miscellaneous reference info for the racer. Includes all hi performance parts for the V6 Chevy. **#S120 . . . $4.95**

ORDER HOTLINE (714) 639-7681 **STEVE SMITH AUTOSPORTS**

Add $1.50 per book for 4th class up to $4.50.   P.O. Box 11631   Santa Ana, CA 92711